T0184724

Bodies, Technologies and Methods

This book examines how different technologies can be used to enhance research methods in the social sciences and humanities.

The boundary between the body and the digital has become increasingly blurred in recent years due to the rise of technologies that capture and reshape our embodied selves. New technologies all too often reflect the attitudes of the privileged white men who dominate the tech sector. This book thus, in part, considers how critical researchers can employ new technologies while challenging some of the problematic assumptions that underpin their design. It also includes a series of case studies that examine the dynamic use of different techniques to explore key questions around the intersection of embodiment and the digital.

With a playful, experimental approach to conducting research today, this book offers new, cutting-edge methods that respond to the potential of different technologies. It will be invaluable reading for undergraduate and post-graduate students of social sciences and humanities to explore ways in which this approach can bring new insights to a range of interdisciplinary research questions.

Phil Jones is a cultural geographer based at the University of Birmingham whose primary interest in developing and deploying innovative methodologies particularly through the use of emerging technologies. He has written extensively on questions around embodiment, creativity and the city.

Routledge Series on Digital Spaces

Series Edited by Chris Lukinbeal and Barney Warf

Digital and physical worlds have become so intertwined they are inseparable. The digital revolution has had enormous impacts on the people, economies, politics, cultures, and places. This book series engages with cutting edge research on the effect of digital technologies on the world(s) we live in, from a variety of perspectives and scales. It welcomes contributions from across the social sciences and humanities that seek to stretch disciplinary boundaries by encompassing new ways of seeing and dealing with digital technologies.

Bodies, Technologies and Methods
Phil Jones

For more information about this series, please visit: https://www.routledge.com/geography/series/RSDS

Bodies, Technologies and Methods

Phil Jones

Routledge
Taylor & Francis Group

LONDON AND NEW YORK

First published 2020
by Routledge
2 Park Square, Milton Park, Abingdon, Oxon OX14 4RN

and by Routledge
605 Third Avenue, New York, NY 10017

First issued in paperback 2021

Routledge is an imprint of the Taylor & Francis Group, an informa business

Publisher's Note
The publisher has gone to great lengths to ensure the quality of this reprint but
points out that some imperfections in the original copies may be apparent.

British Library Cataloguing-in-Publication Data
A catalogue record for this book is available from the British Library

Library of Congress Cataloging-in-Publication Data
Names: Jones, Phil, author.
Title: Bodies, technologies and methods / Phil Jones.
Description: Abingdon, Oxon ; New York, NY : Routledge, 2020. |
Series: Routledge series on digital spaces | Includes bibliographical
references and index.
Identifiers: LCCN 2019053995 (print) | LCCN 2019053996 (ebook)
Subjects: LCSH: Technological innovations–Social aspects. | Technological
innovations–Research–Methodology.
Classification: LCC HM846 .J66 2020 (print) | LCC HM846 (ebook) |
DDC 303.48/3–dc23
LC record available at https://lccn.loc.gov/2019053995
LC ebook record available at https://lccn.loc.gov/2019053996

ISBN 13: 978-1-03-223696-4 (pbk)
ISBN 13: 978-0-367-19583-0 (hbk)

Typeset in Times New Roman
by Taylor & Francis Books

Contents

Figures

Contributors

Author biography

Phil Jones is a cultural geographer who has been based at the University of Birmingham since 2000. His first degrees were in history before moving to geography for a PhD examining historical urban development. Since then his research has roamed around different topics including mobilities, urban sustainability, creative economy and arts practice. The common thread to his work is an interest in methodologies, particularly in how new technologies can be used to enhance approaches to qualitative research. He runs the Playful Methods Lab at the University of Birmingham and has a keen interest in interdisciplinarity, collaborating with colleagues across the humanities, social and physical sciences.

Contributor biography

Tess Osborne is a social geographer based in the Population Research Centre, Faculty of Spatial Sciences at the University of Groningen with an interest in mobile mixed methodologies and embodiment. Prior to joining the University of Groningen, she undertook her PhD research at the University of Birmingham on the use of biosensing technologies for researching memory and heritage spaces. She now works on an ERC-funded project investigating indoor and outdoor mobilities of older adults in the UK, the Netherlands and India.

Acknowledgements

This book draws together ideas that I have been developing for over a decade meaning that there are a great many people to mention who have helped along the way.

Colleagues at the University of Birmingham have played a major role in helping me to sharpen up my thinking as well as providing moral and intellectual support over the years. A big thank you to Dominique Moran, Lloyd Jenkins, Lesley Batty, Julian Clark, John Round, Jon Oldfield, Lauren Andres, Jessica Pykett, Bill Bloss, David Hannah, Jon Sadler and Susannah Thorpe alongside many others.

I have been fortunate to have a number of fantastic young scholars pass through my Playful Methods Lab, including Saskia Warren, Arshad Isakjee, Tom Disney and Colin Lorne. Tess Osborne, a former member of the lab who is a co-author on one of the chapters here, needs to be singled out for particular thanks having reignited my enthusiasm for tech and research.

I have had a number of great interdisciplinary collaborators and co-conspirators over the years who have helped expand my thinking beyond the bounds of human geography. Just to mention a few from a very long list: Antonia Layard, Chris Speed, Paul Long, Dave O'Brien, Chris Jam, Jodi Sita, Anna Goulding, Stu Denoon-Stevens, Richard Clay, Ceri Morgan, Sigríður Kristjánsdóttir, Emily Warner, Dan Burwood, Neil Macdonald among many others. A hat tip must also be made to friends and colleagues who are part of the Digital Geographies Research Group of the Royal Geographical Society – a pleasure to work with so many digital enthusiasts!

Thanks to a Universitas 21 grant I was able to spend time as a visiting fellow in the School of Geography at the University of Melbourne in 2018. It was during this period that I finally found time to put together the proposal for this book and particular thanks go to Lesley Head for kindly agreeing to host me within the School as well as to David Bissell and Tim Edensor for some really stimulating conversations which helped to crystallise some of the ideas I present here.

My ballroom partner Elizabeth Hayes and friends from the Dance Workshop have been a great sounding board over the years and provide a much needed reality check whenever I'm in danger of disappearing inside my own navel.

Finally, of course, my parents and brother have been a constant source of support, common sense and the occasional kick up the backside along the way, all of which have been much appreciated.

Acronyms

AFR	Automated facial recognition
AI	Artificial intelligence
AR	Augmented reality
CT	Computed tomography
EAT	Experiments in Art and Technology
EDA	Electrodermal activity
EEG	Electroencephalogram
fMRI	Functional magnetic resonance imaging
FR	Facial recognition
GCHQ	Government Communications Headquarters
GIS	Geographic information system
GPS	Global positioning system
MET	Mobile eye tracking
MMOG	Massively multiplayer online game
NFC	Near field communication
IoT	Internet of Things
RFID	Radio frequency identification
VR	Virtual reality

1 Introduction

Playing with methods

My father is a tremendously generous man. When it came to buying his children the UK's must-have toy in the early 1980s, he did not flinch at getting my brother Chris and I a brand new ZX Spectrum computer. My mother definitely appreciated the peace as her two rambunctious sons took a break from running around and playfighting to be quietly distracted by this new device. The computer was nominally educational, in that you could learn to program it, but the real joy was in the games that today are guaranteed to make a certain kind of middle-aged British person misty eyed with nostalgia. *Chuckie Egg, Hungry Horace, Skool Daze, Jet Pac* and, king of them all, *Jet Set Willy* were the touchstones of my youth. As we became teenagers, Chris moved on to interests in science, flying and girls, while I retained the fascination with computing and technology even as I went on to study in the humanities and later the social sciences.

Today, Chris laughs at me whenever I justify the purchase of a new device by saying that it is 'for work' knowing that this is only ever half true. For all that this book could easily be written as a love letter to technology, however, as a critical social scientist one cannot help but by horrified by some of the ways that technologies have been developed and employed, particularly in the last two decades. There is a lot to be concerned about, from social media companies generating advertising revenue by fostering addictive behaviours, security systems that normalise racial profiling, through to the dungheap of radicalised white nationalism and misogyny that festers within gaming cultures. Many of these issues have been subject to intense academic scrutiny by scholars far more qualified than I and this book is not simply another critical account in this vein. My intention here is instead to concentrate specifically on the methodological implications for critical scholars of this changing technological landscape.

This is a book about methods, but it is not a how-to guide. Again, there are a great many books that can take people through the specifics of approaches to data collection in the social sciences and humanities. These stretch from the relatively generic such as how to undertake an interview or archive work to more specialised techniques such as using R for spatial analysis (Brunsdon and Comber 2015), or the application of Q-method (Watts and Stenner 2012).

My intention here is, however, to zoom out from the step-by-step specifics and instead explore some of the new techniques that we have at our disposal, how these are changing the ways we work, and what the implications of this are for undertaking critically informed, ethical research.

This being said, some of the methods I discuss here are not necessarily all that new in themselves. One of the big changes that has been happening in recent years is a much greater transfer of approaches between disciplines, partly enabled by new technologies that have reduced the price point and complexity associated with employing what had previously been highly specialised techniques. Although my first degrees were in history, I now work as a geographer, which is notoriously a magpie discipline, grabbing approaches and topics from across the humanities, social and physical sciences. This book is not intended to cater solely to a geography audience, however, instead being aimed at critical scholars from a variety of disciplinary backgrounds, with a view to encouraging interdisciplinary thinking and approaches. Exploring methods developed in different disciplines allows us to answer research questions we may not have previously considered. This approach also allows develop new ways of thinking simply because we can gather different kinds of data.

Many of my collaborators working in the humanities squirm when I talk about *data* collection rather than gathering research *materials*. Because I have worked for many years as a social scientist within a physical science department, I have undoubtedly picked up some of the language and mindset of the hard scientist, if for no other reason than protective camouflage. Another example of this comes in the fact that I talk about running a Playful Methods *Lab* because this is the kind of idiom my physical science bosses understand, even if they have no real idea what I do. In practice the 'lab' is merely the branding for the team of PhD students and postdocs that I've worked with over the years plus a cupboard filled with different bits of tech that I keep in my office.

As part of this first chapter, I will introduce my core approach of *playing* with methods, which might, less charitably, be described as messing about with different techniques to see if any of them generate useful research materials. Having the freedom to play and experiment is, in essence, is what this book is about. I turn first, however, to examine why questions of embodiment matter when considering methods shaped by new technologies.

Whose bodies?

I have always found teaching to be a crucial medium through which to pressure-test ideas that I generate through my research – if it strikes a chord in class, then I usually feel that I am on to something. As a result, I will occasionally refer to working with students throughout this volume, even though it is not an undergraduate textbook. One of the key advantages of working in the university sector is the potential to co-construct research and teaching, with a blurred boundary between the two being productive for both.

For many years I have taught modules examining questions around embodiment, usually attracting small but very devoted groups of undergraduates. My intention with these modules has always been to ground challenging theoretical concepts in our everyday lived realities, demonstrating the *relevance* of those theories beyond mere intellectual exercises. To give a practical example, Foucault's concept of *discipline* generated through panoptic surveillance comes to life when students start making connections to their everyday experiences, not least how the university monitors their presence in class and within online learning platforms. Judith Butler's ideas on language and labelling shaping who we are seems quite abstract in isolation. When reflecting on how female students alter their behaviour and dress in anticipation of sexist abuse being shouted at them in the street, however, Butler's ideas suddenly become both tangible and powerful.

In recent years, questions around the technological have become a greater part of my teaching as I discuss ideas around embodiment being co-constructed through technology. Of course, this is not a particularly new insight. Donna Haraway (1985) was using the metaphor of the cyborg in the 1980s to talk about the blurred boundaries between humans, animals and machines. Don Idhe, meanwhile, has been writing on the philosophy of technology since the 1970s and his work on postphenomenology has examined the shaping of lifeworlds through the intersection of embodiment and technology (Idhe 1990).

Fascinating though these theoretical debates are, this is not a theory-led book. Nonetheless, much of the work on embodiment explicitly draws on feminist theory and these perspectives are critical to the ideas that I discuss throughout the chapters that follow. In examining these debates, however, I am acutely aware of my status as a middle-aged, middle-class, straight, white, cisgender man and how these characteristics give me a position of power and privilege. Indeed, many people with whom I share these traits assume that the white male perspective is the default, rather than merely one view among many. This is a particular problem within the tech sector, which is dominated by people who, frankly, look a lot like me. Gendered, raced and heteronormative assumptions run rife in Silicon Valley and this creates significant issues with how embodiment is conceptualised and shaped by new technologies. Reflecting on the inventions of the tech sector, Tabitha Goldstaub (2017) has used the analogy of crash test dummies: because these were originally designed around male physiology, car designs were less protective of women. The same kind of problems cut across a large number of issues in how technology is designed, for example in the way that many user interfaces take no account of intersectional identities (Schlesinger et al. 2017). Fundamentally, many of the assumptions built into the tech sector are predicated on the bodies engaging with it being able, white, male and otherwise privileged.

Katherine Losse's (2012) memoir about her time working for Facebook captures some of the heady atmosphere of being close to the heart of a business on its way to global domination. In *The Boy Kings*, Losse also highlights many of the gender inequalities within the sector, the lack of women in the higher-paid engineering jobs, the casual harassment of female employees, treating them as objects, categorised as either 'pretty' or 'witty'. A raft of similar memoirs and accounts have followed, including Emily Chang's (2018) *Brotopia* which explored toxic male cultures in Silicon Valley, the sex- and drug-fuelled parties and the ways that women were locked out from the personal networks that drive investment and product development. Beyond these more personal accounts, the deep sexism facing women working in the tech sector has been captured in the report *Elephant in the Valley* (Vassallo et al. 2016), which surveyed 200 leading women in the industry. Of these, 84% had been told they were 'too aggressive' in the office, two-thirds reported being excluded from key workplace events because of their gender and 60% having experienced unwanted sexual advances in the workplace, the majority of which were by a superior.

When women are not taken seriously as colleagues this has a series of knock-on effects. Certain behaviours and priorities may not be called out for being sexist simply because there are no women in the room who can do so. Even a company like Apple, which likes to market itself on being forward thinking, did not undertake a diversity report about itself until as late as 2014. This review revealed that 70% of its employees were male and more than half of its workers in leadership positions were white (Lowensohn 2014). In that same year the company released its first Apple HealthKit, which was sold as allowing people to 'monitor all of your metrics that you're most interested in' and yet did not include a menstrual cycle tracker (Duhaime-Ross 2014). The company faced major criticism for this and added period tracking in the iOS9 update a year later (Perez 2015) but the fact that it simply did not *occur* to anyone that this might be a useful thing to include from the start is symbolic of the kind of thinking that arises when there are no female voices present in strategy meetings. We see something similar in the fact that the major digital assistants (Alexa, Siri, Google Assistant, Cortana) are gendered female by default, the underlying assumption being that an assistant's job is women's work. A UNESCO report picked up on this issue noting that until changes made in early 2019, Siri responded to the comment 'Hey Siri you're a bitch' by saying 'I'd blush if I could', with digital assistants being a model of submissiveness in the face of sexist abuse (UNESCO 2019).

Since around 2015 there have been *some* moves to clean house within the tech sector, although not a great deal has changed in terms of hiring practices. Nonetheless, there have been some high profile actions, such as Travis Kalanick being fired as CEO of taxi company Uber once it became clear that he had presided over an office culture of rampant harassment and frat-house misbehaviour (Isaac 2017). Unsurprisingly, perhaps, the attempt to reform led to a pushback from some men within the sector who saw themselves as now being the victim of discrimination. In the summer of 2017, for example, James Damore, an engineer at Google, published a memo filled with pseudo-scientific language stating that men were biologically inclined to be better at writing code than women. He argued this meant Google's diversity policies were fundamentally flawed because women were simply not as capable as men (claims which have been carefully debunked by, among others, Campbell 2017). Unsurprisingly, once his memo became public, Google fired him. He instantly became a heroic martyr to those supporting the alt-right, who claimed that he was sacrificed for daring to speak the truth about female inferiority. Indeed, Google continues to be criticised in the right-wing press for supposedly discriminating against those with conservative political views (Ghaffary 2019).

The tech sector undoubtedly has an uncomfortable relationship with embodiment and particularly with female embodiment. Facebook's Mark Zuckerberg began his journey toward world domination in 2003 with tool called FaceMash which used stolen images of female students and asked people to rate their attractiveness. At about the same time Angela McRobbie (2004) was talking about the rise of post-feminism within media discourses, where the language of empowerment was being used as a cover for moves that undermined the gains to female equality that had been made in the 1970s and 1980s. The post-feminist con trick has thus run in parallel to the rise of social media since the early noughties. Unsurprisingly, there has been a mountain of commentary and academic work exploring how gender discrimination is being played out in new types of media, with particular concerns around mental health. Being acutely aware of the tech sector's structural sexism, therefore, sends up red flags when examining the ways that we are encouraged to present ourselves online and particularly questions around photography and body image (among many others, see for example Tiggemann and Slater 2013, Fardouly et al. 2017).

It would be a little too easy, however, to simply position technology as a bogeyman producing all of our social ills. Throughout this book I will examine positive changes enabled by new technologies, as well as some of the genuinely unpleasant social phenomena that have been driven by how the tech sector has evolved. We also need to be aware of powerful actors attempting to shift the blame for wider structural problems onto new technologies. We see this, for example, in discourses of moral outrage about the corrupting influences of violent video games on young people. Following mass shootings that took place in Dayton and El Paso in summer 2019, President Donald Trump and a host of right-wing commentators blamed, among other things, violent video games rather than the widespread availability of firearms (Hernandez 2019). Those of us on the political left can all too easily fall into the same kind of intellectual trap by suggesting that the tech sector's diversity problem makes all of its products irredeemably problematic.

The important point to reflect on here is that technology is not neutral – it is a product of the imaginations and flaws of the people who make it and the wider society that shapes and is shaped by them. It is a fair assumption that, for the foreseeable future, the products of the tech sector will certainly be more orientated to positivist and masculinist perspectives with an emphasis on quantification. As critical scholars, however, we should avoid the temptation to dismiss the research techniques enabled by new technologies as being tainted, rather than positively engaging with the possibilities that they offer. We can attempt to work around or subvert the underlying assumptions on which different technologies were built, creating social aware research aimed at making positive changes in the world around us. Indeed, refusing to engage all too easily becomes a cop-out for not exploring new ways of doing research. New methods may be challenging to our existing assumptions, difficult to operationalise or simply *different* to what we are used to doing but those are poor reasons for not trying them. Throughout this volume, I celebrate some of the work being undertaken by critical scholars who take technology in new directions within their research.

Playing with methods

All research methods are a compromise, generating data that is situated and partial, only allowing us to see part of the picture. For most of the last decade I have been interested in developing new research methods or new applications of existing techniques. People working on advancing research methods often face the critique that we prioritise innovation for its own sake and thus dismiss robust and well-established techniques. To avoid any confusion, therefore, I will clearly state here that standard research techniques are incredibly valuable. For the vast majority of projects, the common methods used in our disciplines will be the most appropriate. My guiding philosophy as a methods researcher is to find the simplest solution that generates the data (or materials if you prefer) that are needed to answer your research questions. For me, the point of methodological research is to try different approaches in order to examine the advantages of a technique, its compromises and how it can be deployed in new contexts. This may mean doing things in an overly complicated manner in order to work out what would be the optimal approach in most circumstances. To give an example, many years ago I worked on a project that used walking interviews as a technique. Among other things that we experimented with were a variety of different microphones and ways of recording the conversations with our research subjects. Ultimately, however, we found that putting a cheap digital Dictaphone on a string around the neck of our participants gave us the optimal balance between useable audio quality and simplicity of operation in the field.

I tend to talk about this experimentation as being part of a 'playful methods' approach – messing about with things until you hit on a good solution. When talking about 'play' within methodology, there is, however, sometimes an assumption that you are talking about research with children. There is, of course, some really interesting work on cognitive theories of play and human development (Takhvar 1988). Likewise, while I would not dismiss the fantastic work happening in childhood studies, I argue that play and experimentation (in a non-scientific sense) are not simply restricted to young people. Indeed, part of the point of working in the academic sector is that despite the increasing intrusion of neoliberal management (Cannella and Koro-Ljungberg 2017) there is still a huge degree of freedom to simply *try things out* which may not have an immediate pay-off.

For me then, playfulness is an approach to research methods that tries new techniques in order to develop innovative insights into a problem. This comes with the risk, however, that such experimentation will not deliver methods that give a significant improvement over conventional approaches. This being said, failure is an important part of research which we do not always capture as well as we might (Harrowell et al. 2018). Indeed, this is a significant problem particularly within the medical sciences where there have traditionally been very few publication routes for *negative data* (Hayes and Hunter 2012). As researchers, if we do not tell people about the things that have not worked, there is a risk that other people will waste time making the same mistakes. Indeed, I have previously published an account of how my own failings in project management had a significant impact on the ability of a research team to deliver meaningful findings, but also took the research in unexpectedly productive directions (Jones and Evans 2011). As a general rule, however, the things that go wrong do not always make their way into publications. Throughout this book I have given examples of things I have done that have not worked, ideas that were never pursued and projects where I have been unhappy with the outcome. The broader point is that failure can be very productive in making discoveries; more playful or experimental approaches are built on the idea that the first thing we try might not work.

There is, however, a very important issue here relating to power and privilege. I was four years post-PhD and in a permanent post before I really started to try out riskier projects. The lower rungs of the university sector are increasingly characterised by a high degree of precarity (Ivancheva 2015). Not everyone has the luxury to try out projects that may not deliver the kinds of outcomes needed to maintain a foothold in the sector. There are significant parallels here to the Silicon Valley cliché of 'failing upwards'. One of the great myths within start-up culture is that entrepreneurs learn more from failure than from running a successful business, which means someone whose first company collapsed can still be attractive to venture capitalists seeking to invest in new ideas. What is left out of this narrative is that the entrepreneurs who do this are overwhelmingly white, male and with social and educational capital that puts them into personal networks with those angel investors (Losse 2016). Research by digitalundivided (2018) estimates that from 2009 to 2017 start-ups led by women of colour secured just 0.0006% of the total tech venture capital funding in the US.

Having the confidence to fail, knowing that the consequences will not leave you destitute and that you will almost certainly get to try again reflects (wealthy, white, male) privilege not only in the tech sector but also in the academy. Many researchers do not have this luxury, but it is my hope that this book highlights some of the possibilities that are opened when engaging with different approaches to research and may give colleagues some intellectual cover from the critiques of sceptical managers as well as the confidence to simply have a go. Broadly, I would suggest that there are two routes to take with a more playful approach and that these carry different degrees of risk.

The first is in discovering a new technique, learning how to use it and directly applying it your own research questions. The second is in learning about that technique in order to have more informed conversations with potential collaborators who are expert in its use. Both approaches see researchers often stepping beyond their own disciplines, but not necessarily fully immersing themselves in another, thus working in a liminal state between disciplinary boundaries. The active learning that comes from playing with someone else's methodological toolkit can generate new ideas and new insights within the research problems you are working on. It can also feed back into the continued refinement of those techniques by employing them in ways that their designers did not anticipate.

Developing new techniques and remixing older ones is not for everyone. Indeed, there is nothing inherently superior in using a novel technique when an existing one will do the job just as well if not better. The inherent biases of the tech sector mean that some of its products come with a particular way of thinking which can in turn erect quite high barriers to their use. This can restrict the opportunities for some people to engage with the methodological possibilities that these products present. In teaching IT classes over the years, I have witnessed many students (male and female) becoming profoundly frustrated because they just don't *get* the way of thinking that underpins, for example, using a spreadsheet. Again, there is a significant gender issue here as generations of young women have been told, in ways both subtle and not, that working with computing and technology is *not for them* (Vitores and Gil-Juárez 2016). This erodes the confidence necessary to try things out and figure out why something you are attempting to do is not working. In turn, this means narrowing the group of people who are willing and able to adopt a more playful approach to using technologies which of course means narrowing the perspectives brought to this type of work.

I am acutely aware that I am somewhat nerdy and, as the story at the top of this chapter indicates, I have been incredibly fortunate in being supported to develop my enthusiasm for technology from a very early age. Not everyone finds this sort of thing fun, which is one of the other reasons I like to emphasise *play* in my approach to working with new methods. Positioning tech as something that can be *played* with, lowers the stakes for people trying out new devices and approaches for the first time. This in turn hopefully helps to overcome the sense of shame and abjection some people feel when they are struggling with a new technique. Getting things wrong is just part of the game.

The structure of this book

The theme that runs throughout this book is that bodies, technologies and methods are all socially constructed, products of who we are and the society we live in. In the following chapters I take a variety of topics in turn, examining their implications for critical scholars and what they mean for the design of research projects.

The choice of how to structure a book like this is always going to be a compromise and there are common threads that cut across these topics. One of the most significant of these cross-cutting themes is the focus of Chapter 2 which examines questions around privacy. There has long been a temptation to reduce people to datapoints, which can then be searched and queried to examine patterns of behaviour – indeed, this is one of the things that makes critical scholars so uncomfortable about quantitative methods. The landscape of the post-noughties tech sector pushes this one stage further by tying those data points directly to an individual, with anonymity being seen as a constraint rather than a basic requirement of good research.

If enough of us can be persuaded to trade privacy for convenience, then a lack of privacy becomes the default option, leaving vulnerable those who really need it. At the same time, a discourse of public safety has been deployed to expand surveillance powers by making claims that people who want privacy have something untoward to hide; this discourse justifies the invention of new technologies that further erode privacy. For all that this might make us uncomfortable, however, the undermining of privacy has actually given researchers some interesting tools which *could* be employed as part of socially valuable research projects. We therefore end up with ethical dilemmas to consider, such as whether the projects we could undertake using, say, facial recognition cameras, are worth the horrendous risk they pose to the privacy and security of participants.

Chapter 2 also considers the role of algorithmic surveillance in monitoring and sorting our lives. These processes reduce people to crude simplifications within a dataset – our likelihood to buy a particular product, the risk we pose to state security – rather than being complex, embodied individuals. In Chapter 3 this idea of reducing our lives into simplified data is taken a step further by considering how bodies themselves are being quantified. Over the last two decades a range of technologies have become available at significantly reduced cost that make it much easier to examine different aspects of our physiology and movement. This chapter is co-authored with one of my former PhD students reflecting the fact that some of this material emerged from her doctoral work, particularly the section examining how measures used by those working in psychology are starting to find their way into other areas of research. One of the most interesting potential applications of physiological measures such as skin temperature, electrodermal activation and eye tracking is in gaining insights into emotional state and decision-making.

There are, however, some significant reasons for critical scholars to be cautious about the application of these techniques. The first is a practical concern: just because physiological measuring devices are now comparatively cheap and easy to use does not mean that the data they generate can be unambiguously understood by non-experts. There is, therefore, considerable merit to collaborating with researchers more familiar with such techniques when designing a research project using these technologies. The other issue is that these measures are highly reductive. Thus, when measurements are taken outside a controlled lab context it can be very difficult to identify what stimulus has provoked a particular physiological response. It can be very productive, therefore, to bring measures of the bodily together with more conventional qualitative techniques. Mixed methods approaches that include measures of physiological response can be used to gain a more rounded understanding of why participants have reacted in a particular way.

Chapter 4 examines how videogames have moved beyond the cliché of being objects of fascination for teenage boys. The gaming sector has become a multibillion-dollar global industry producing complex texts that are ripe for analysis in their own right while also being valuable tools for undertaking research on a variety of topics. The chapter reflects on how games as *playable* texts differ from conventional sources for media analysis. Despite appearing to be a visual medium, gaming is in fact multisensory and this gives opportunities to investigate a range of embodied responses to the virtual worlds that players encounter.

As with the wider tech sector, gaming has traditionally had a problem with representations of gender, sexuality and ethnicity although this is slowly starting to change. Nonetheless, this does have implications for *who* is able to carry out research in this area and how they do it, particularly in multiplayer scenarios. The complexity and sheer detail of contemporary games often surprises people seeing them for the first time and there is much work to be done in examining games as landscapes, how they are consumed and *explored* although this can raise questions from critical scholars using postcolonial theory. Building toward the following chapter, this section also examines the ways that games technologies are being directly employed by scholars within research projects. Freely available games engines allow us to become more creative in our research practice by building our own virtual worlds in which we can examine a range of issues with our participants.

Chapter 5 develops this theme to explore wider questions around creative practice. The chapter covers a somewhat broader area than conventional arts-based research by moving beyond the need to produce work with any overtly aesthetic ambition. Bodies are messy, fleshy objects and this has significant implications for how creative engagements with different technologies can be undertaken. A variety of technologies are available that allow researchers to become involved in the process of creating new products and experiences, even if it is simply a case of playing around in order to have a more meaningful conversation with a specialist working in that area. Projects involving VR and AR technologies, for example, creatively rewire our understanding of the spaces around us by giving varying degrees of immersion in virtual worlds. Creative practices also offer the opportunity for researchers to take a more activist approach, using our projects to directly change elements of the world around us, hopefully for the better. Of course, there is a need to be cautious here, particularly in imposing our preferred visions onto communities who will nominally benefit from this activity. Indeed, when using creative approaches with communities, co-construction of the research objectives and outcomes is often the best approach.

Chapter 6 considers how new technologies are changing the nature of our mobilities and the ways in which we use maps to represent the world around us. Research drawing on the 'mobilities paradigm' has been interested from its inception in the potential of different technologies to shape our practices of moving around the world. Methodologically, work in this area has drawn heavily on (auto)ethnographic accounts and video. The chapter examines how video methods have been changed by the availability of high quality, increasingly portable and cheaper cameras while also exploring how other technologies can add valuable insights to our studies of everyday mobilities.

The second half of the chapter looks at how mapping technologies have been radically changed by the rise of online map services and location-aware, GPS-enabled smartphones. Maps come from positions of power, but a very large amount of work has been done by critical scholars using geographical information systems (GIS) to destabilise and subvert this power. Indeed, far from attempting to control and dominate communities, a great deal of contemporary scholarship using maps seeks to highlight spatial injustice and suggest potential solutions to social problems. Meanwhile, anyone who owns a smartphone has now become a kind of cartographer, because these devices create spatially referenced information about different aspects of our lives. This represents a great opportunity for critical scholars to work with communities to create new representations of their neighbourhoods. These can be used to counter official, top-down understandings of those locations and the people who live there.

In the final chapter of the book, I examine some of the broader lessons for scholars using new technologies to undertake critically informed research and next steps for those wanting to adopt a more playful approach to their own projects. These kinds of approaches to research are unavoidably shaped by the position of those undertaking them, as well as that of the people who designed the technologies we deploy. Nonetheless, the book ends on an optimistic note, suggesting that it is our responsibility as critical scholars to use these technologies to try to make the world a better place, even if the intention of their designers can sometimes be seen as problematic.

2 Privacy, transparency and ethical research methods

Introduction

As I grew up during the tail end of the Cold War in the 1980s, there were simple narratives played out in popular media about how malevolent the Soviet Union was. The 'evil empire', we were told, was propped up by the secret police, spying on citizens to make sure they never spoke or acted against the regime. East Germany's Stasi thus became the real world manifestation of Orwell's Thought Police (Rodden 1987). Within this discourse, privacy was positioned as a marker of freedom within the capitalist West.

Since the collapse of the Soviet Union it has become abundantly clear that the desire to spy on the lives of citizens is far from an exclusively communist trait. In 1999 Scott McNealy, the CEO of Sun Microsystems, told a group of journalists at a product launch 'You have zero privacy anyway ... Get over it' (Sprenger 1999). Sun was later bought out by rival firm Oracle and, with no small irony, its Menlo Park headquarters sold to Facebook in 2011. There has been no shortage of privacy scandals in recent years that have reaffirmed McNealy's somewhat depressing claim, from Edward Snowden's disclosure of mass surveillance by the US government to Facebook's admission that it had allowed Cambridge Analytica access to a raft of personal data which influenced both the election of Donald Trump and the Brexit referendum. While the Cambridge Analytica revelations meant that Facebook suffered a small drop in users (and enormous drop in stock price) in the summer of 2018, its core business model of selling information about its users to advertisers remains both intact and highly profitable. Indeed, in response a record $5bn fine levied by the Federal Trade Commission in 2019 for the Cambridge Analytica affair, Facebook's share price actually *rose* because traders were reassured that the company's business model was not going to be threatened with government regulation (Kang 2019). Relative to Facebook's turnover, the fine was minuscule.

The availability of personal data gathered about millions of people via their interactions with different technologies has a tremendously seductive power for researchers. To take just one example, anyone can freely search a 1% sample of posts for the last week on Twitter, while those willing to pay for 'firehose' access, can search through every tweet posted by every user since 2006. Amazing research has been undertaken using these 'big' datasets to look at a range of issues from real time monitoring of disease (Lee et al. 2013), crowd and traffic management (Arribas-Bel et al. 2015), monitoring emotional response to US presidential debates (Wang et al. 2012), to residents' use of urban green space (Roberts et al. 2019). The danger, of course, is that in the excitement about the potential of what can be done with such datasets, we lose sight of the individuals who generate those data for us.

Data scientists are far from uncritical about these issues and have themselves been understandably concerned about the ways in which new technologies allow ever greater intrusion into the lives of individuals (Smith et al. 2012). As discussed in the previous chapter, feminist scholars have meanwhile lamented the overrepresentation of men within the tech sector and the often toxic cultures that underpin the production of new tools and techniques (Wachter-Boettcher 2017). This is not necessarily a question of malice or evil intent, but more a failure to think through the consequences of developing and deploying technologies in ways that are ethically questionable, particularly in relation to privacy.

In this chapter I explore some of the ways in which questions of privacy have become all pervasive when considering the technologies that have come to dominate our lives over the last two decades. The challenge is to examine the trade-offs and compromises that come with critical research that employs technologies originally created for ethically questionable purposes. It is intellectually lazy to merely dismiss the research methods that these technologies enable as being positivist and therefore *bad*. As I argue throughout this book, instead we have a duty to challenge uncritically positivist approaches by exploring alternative ways of using technologies and subverting the cultural assumptions that underpin them.

The power of the intimate

When Foucault (1977) wrote about the nature of surveillance in the model prison, it was as part of an exploration of power and control over the body; the erosion of privacy gives those who have access to personal information power over the individual. Yeung (2018) talks about the importance of privacy as a collective good, which protects the individual and allows them to establish different identities and remake themselves on their own terms. Her concern is that:

> If a sufficiently large critical mass of individuals decide, independently, that they are willing to waive their rights to privacy in return for convenience and efficiency, then the resulting social structure cannot sustain a "privacy commons" and cannot therefore provide a zone of privacy for its members, including those who choose not to waive their individual rights to privacy.
>
> (Yeung 2018, 518)

The problem for citizens today is the extent to which there are realistic alternatives to ceding so much power over their own lives to technology companies whose business models rely on hoovering up information about us. Tech today is no longer a sector one can choose to engage with or not; it is a fundamental underpinning to all aspects of everyday life. This creates a tremendously unequal power relationship between tech companies and their users; terms and conditions are imposed on a take it or leave it basis, knowing that users have little choice. The result is that the privacy commons has been severely eroded.

The European Union has been in the vanguard of attempts to give power back to the individual in an age of technology-led mass surveillance. A 2014 ruling by the EU Court, for example, required search engines to enforce a 'right to be forgotten' upon request by removing search results about an individual where they would be deemed 'inaccurate, inadequate, irrelevant or excessive' (European Commission 2014). Indeed, stringent rules on how personal data about EU citizens are stored and used by companies, governments and third sector organisations were subsequently introduced in the 2018 General Data Protection Regulation, which strengthened the right of erasure and forbade the re-use of personal data for anything other than the original purpose for which it was gathered (European Commission 2018). Arguably, this brings the private sector more into line with the kinds of restrictions that university researchers are familiar with through processes of ethical review. Being clear with participants about what kinds of data are being collected about them and not using those data for purposes other than those for which the participants have given their explicit, informed consent has long been at the heart of scholarly research.

The invasion of new technologies into every aspect of our lives goes well beyond the information we share on social media. The paradox for researchers is that these data are simultaneously intimate and disembodied. The individual is reduced to a series of data points – the websites visited, the products bought, the stories 'liked'. Drilling down into these data can reveal many intimate details about the individual. We can, for example, predict various demographic qualities of an individual just from analysing a sample of their anonymised tweets (Volkova et al. 2015). Being able to predict an individual Twitter user's salary with 60% confidence is, no doubt, an invasion of privacy. This is, however, pretty thin stuff compared to the messy embodied reality of that person's life. Identifying very broad categories of demographic characteristics does not say very much about an individual's *personhood* in a complex world. Nonetheless, the insights we can glean into people's lives through these techniques are unambiguously *creepy*.

Sticking with Twitter as an example, I have asked students to undertake an exercise where they use a basic tweet harvesting tool embedded in a Google Docs spreadsheet.[1] This does not require any programming skill and allows students to easily search the 1% sample dataset for the past week. A small proportion of tweets contain geotagging information, meaning that the GPS location of the mobile device used to send the tweet is also recorded. This allows you to map GPS data for individual users, through which you can start to get insights into their daily movements – where they live, where they work, how often they travel beyond their immediate locality – as well as the things that interest them sufficiently to tweet about.

The response to this exercise, which is intended to get students thinking about questions around privacy, is usually how *uncomfortable* this makes them. When you sign up for Twitter, you give consent for the data you generate to be made public and thus shared and analysed in ways that perhaps you never expected. As a researcher there is an unpleasantly voyeuristic element to work of this kind, giving a window into someone's life whom you will never meet and who will never know that you are observing them. Again, however, for all the discomforting intimacy of such an encounter, the insights gained about that individual are superficial, reflecting narrow slices of their everyday embodied encounters with the world. This, in turn, raises important questions about our temptation to make judgements about these individuals – who they are and how they behave – based on such a superficial insight into their private lives. As decades of feminist scholarship have warned, research participants should never be reduced to mere objects of study. It is important therefore to examine the kinds of safeguards should be put in place when considering using research techniques based on a simultaneously intimate yet distal understanding of embodiment.

Capturing the bodily

Thus far in this chapter I have focused on social media data, but as the rest of the book explores, a whole range of techniques have become available that allow us to understand embodiment in new ways. Concerns about how technology can compromise bodily privacy are not new however:

> Instantaneous photographs and newspaper enterprise have invaded the sacred precincts of private and domestic life; and numerous mechanical devices threaten to make good the prediction that "what is whispered in the closet shall be proclaimed from the house-tops." For years there has been a feeling that the law must afford some remedy for the unauthorized circulation of portraits of private persons; and the evil of the invasion of privacy by the newspapers, long keenly felt, has been but recently discussed …
> (Warren and Brandeis 1890, 195)

Just as the invention of photography had a transformative effect on how bodies were represented in the 19th century (Rose 2000), so the volume of data being produced today and the speed at which it can be analysed represent a step change in the power we now have to capture the embodied self.

The right to privacy is a right to curate how the individual presents themselves to the world (Westin 1967). One of the main justifications for curtailing this right is in maintaining public safety (Weinstein et al. 2015). Indeed, researchers working on subjects around the fringes of criminal activity struggle with balancing their responsibilities to protect their participants and in reporting illegal activities to protect the wider public. The sticking point is in where the balance lies between these things. A somewhat chilling example of the tension between privacy and public safety is in the increased use of facial recognition (FR) technology.

For many people, the most apparent use of FR today is in automatically curating online photo albums. FR draws upon a broader set of artificial intelligence algorithms that researchers are training to recognise different characteristics within images: is it a portrait or a landscape; what time of day was it taken; does the image contain clouds, cars, people, etc.? As the technology has become more sophisticated so it appears more uncanny, not only recognising the individuals depicted in your photos but also classifying a scene depending on the apparent mood of the people within it based on their facial expressions (e.g. 'happy days'). As researchers, this kind of technology provides us with an opportunity to search through, classify and quantify a large corpus of photographs; this could have a range of potentially interesting applications. Sitting with participants and asking them to talk through an automated analysis of their family photographs might, for example, add an intriguing methodological twist to the kind of work done by Gillian Rose (2010) in examining how families curate and understand their own lives through photography.

FR has considerably more controversial applications than simply categorising family photos, however, and is most commonly used within the security industry. For example, a number of trials have already been undertaken by UK police forces at major public events using automated facial recognition (AFR). The underlying principle is understandably seductive to law enforcement professionals. Video cameras are connected to a database of known criminal suspects and allow for a non-invasive scanning of large crowds of people. Where a potential match occurs, operators can flag this to officers on the ground who can go in and make an arrest.

There are a number of problems here. Scanning everybody in a crowd starts from the presumption that we are all potential criminals. In one of the pilot studies, Leicestershire Police partnered with the Download Festival of 2015; when buying a ticket, buried deep in the terms and conditions, the 90,000 attendees gave their 'consent' to be part of an AFR trial. The organisers even included RFID tags in the Festival wristbands so that they could track every participant as they moved around the different parts of the site. Leicestershire Police were given a warning by the UK's Information Commissioner indicating that the pilot was inappropriate and disproportionate (Martin 2015), but a number of others have taken place since. The Metropolitan Police monitored the Notting Hill Carnival of 2016–2017, barring entry to those that the AFR system identified as known criminals and troublemakers.

Restricting people's access to public space without trial or ability to appeal points the way to a somewhat horrifying future. What makes this even worse is the fact that FR technology being used to make these judgements is *deeply* flawed. South Wales Police have published figures about their pilot work undertaken at major public events. The success rate of the system in correctly identifying known criminals has been very poor – including a UEFA Champions League match where just 173 people were correctly identified out of 2470 flagged up by the system (South Wales Police 2018). Indeed, analysis afterwards showed that the Notting Hill trial correctly identified just two people, neither of whom was a criminal; the Metropolitan Police decided not to repeat the pilot for the 2018 Carnival (Hill 2018).

The issue here is the databases that are used to train the AI algorithms. This has been recognised as a major problem in recent years, as the capacity of AIs to identify certain characteristics in an image depends very much on the kinds of data they are fed to 'learn' from. In 2015 Google Photos hit the headlines when a black software engineer Jacky Alciné pointed out that photos of he and his friends were being automatically classified as 'gorillas'. Google promised to fix this, but three years later it was clear this was still a problem, that Google had 'solved' simply by preventing the algorithm from labelling *any* image as containing a gorilla (or, indeed, chimpanzee, chimp or monkey, Simonite 2018).

In part, this straightforwardly reflects unconscious bias in the tech sector. A lot of the algorithms underpinning machine learning are unintentionally racist and sexist because of how the corpus of data from which they learn has been constructed (Noble 2018). As a result, FR technologies do not work very effectively for non-white people meaning that AFR is a fairly dangerous technology to be rolling out in a sector as racially-charged as law enforcement. By putting the onus on the individual to defend themselves against an incorrect identification, AFR reverses the fundamental principle of being innocent until proven guilty. This is not just happening in policing, but in databases being built up by the private sector whereby you can be prevented from accessing certain shops and malls because a privately operated AFR system has identified you as a 'known' troublemaker – with no right of appeal against this automated judgement (Lewinski et al. 2016).

Where does this leave us as critical researchers? FR technologies exist and are being used by scholars often in both interesting and important ways – for example in detecting predisposition to genetic disorders (Mok and Chung 2017), tackling child sexual exploitation (Ferguson 2015) and tracking missing persons (Parks and Monson 2018). Researchers working in this area are not unaware of the privacy implications of using this technology. Parks and Monson (2017), for example have demonstrated the potential risk that patient identities could be revealed from computed tomography (CT) scans. New software has made it possible to reconstruct a person's facial features from CT scans which can then be successfully matched to a database of actual patient photographs. Thus, there are potential privacy implications if nominally anonymous CT scan images are made publicly available.

A slew of patents exist for realtime tracking of individuals moving through spaces using AFR. This technology is appealing to managers for the convenience that it offers – for example, automatically creating attendance registers in schools, universities and workplaces (Mehta et al. 2017). Critical scholars, understandably, may flinch at this level of panoptic monitoring, but it does allow us to ask questions about whether this kind of system could be put to a more ethically responsible purpose. Imagine a scenario where a researcher is working with a community group that acts as 'friends of ...' a local park in a relatively deprived neighbourhood. In lobbying for additional funds to continue their important work, they need information on how many people use the park, how many repeat visits they get from particular individuals, the demographic and ethnic makeup of their visitor cohort, perhaps even visitors' emotional state. Theoretically an FR system linked to CCTV cameras at the entrance to the park could help produce this data, perhaps not with 100% accuracy, but because of the scale of the monitoring, it should generate a huge amount of data with sufficient robustness to allow the community group to build a case for new investment.

Such a research project would be inherently risky. Of course, the data would be anonymised but, to work, images of individuals would need to be stored and repeat visits monitored and documented. This would place individuals at risk as a database would exist of their routines which could leak or be hacked with the potential to cause danger or embarrassment – from empowering a stalker to revealing an illicit rendezvous. The counterargument might be that if you could count unique park users on different days, connected to the time of year and prevailing weather you gain some very interesting insights – at scale – about park usage. While recognising emotional state from such imagery might not be entirely reliable, aggregated across thousands of users, one could potentially be able to demonstrate that, for example, people had more positive emotions upon leaving the park than when they came in. Again, this kind of finding, based on an enormous dataset, would be pretty compelling evidence when lobbying policymakers to invest in that space.

Off-the-shelf systems currently available for doing this kind of monitoring work are too expensive for community use, but they are already filtering into schools for monitoring pupils. Thus, while the scenario discussed above seems hypothetical at the moment, as the price point of this technology falls, so community groups and their researcher allies could find themselves attempting to balance these ethical dilemmas. At a personal level, I'm still inclined to feel that the risks of using FR, even with appropriate safeguards to protect the privacy of research subjects, are too high. Of course, it is easy to adopt such a principled position at present while the tech remains just out of reach for everyday research.

Mobile bodies and wearables

Unlike FR, the price point and usability of wearable technologies today places them well within the budget of even small research projects. Around 2013 there was considerable excitement in the tech sector in the belief that the wearables market would be the next big thing. Apple, whose growth was highly dependent on a single product, the iPhone, was keen to diversify its portfolio. Somewhat late to the party, the company released its Series 1 Apple Watch in 2015. Even with the might of the Apple marketing machine and fanbase behind it, however, the wearables market has singularly failed to become the next big thing, remaining stubbornly niche.

In some ways, it is not hard to see why this is the case. Is it really that much more convenient to tell the time, check messages or make a contactless payment using a device on your wrist compared to simply pulling out your phone? Conventional watches, meanwhile, do not need to be replaced every few years as software support is withdrawn. The one area in which wearable devices excel is where individuals are interested in monitoring their own body for reasons of health and fitness. This is, however, always going to be a minority interest.

Nonetheless, the fitness wearables market has developed rapidly. Basic devices use gyroscopes to give relatively accurate estimates of the number of steps a user takes each day. More sophisticated algorithms can use these motion sensors to calculate the number of calories burned in a day, the quality of a golf swing, the amount of restful sleep an individual gets in an evening and so forth. Some researchers have even demonstrated how foot strike and other measures of gait can be calculated through this technique (Félix et al. 2017). More expensive smartwatches include GPS tracking and there are devices available with sensors that can record heart rate, blood volume pulse, electro dermal activation and skin temperature. There are even techniques to measure sweat rate and analyse its chemical composition in real time, although here there remain significant technical barriers to these being adopted by mainstream devices (Heikenfeld 2016).

The methodological possibilities of working with wearables are discussed further in the next chapter. Here, however, we need to reflect on the implications for individual privacy and safety. The first thing to highlight is that many wearables connect to a cloud-based service where the manufacturer offers data analytics. Many devices sync seamlessly with the cloud via a smartphone app, meaning that the user never has to get involved with remembering to download the data themselves. Instead, users are presented with nice graphs and tables that the manufacturer has generated from the data, showing your progress across time and, in some cases, in comparison to other users.

One of the issues here is that many of the manufacturers involved make users sign away their rights to the data they generate. In essence, by using the device, you agree that all the traces of your body that it records are the property of the manufacturer. This removes the user's control over that data, trading convenience for intellectual property rights. As the various scandals relating to Facebook demonstrate, however, this means that personal data can be used in ways that people do not expect. Indeed, some of the debates within computer science about wearables and individual privacy focus on *digital literacy*, educating users around their understanding of how their data can be (re)analysed and used (Taylor et al. 2017). Of course, focusing solely on education responsibilises the user rather than the provider of these technologies.

Strava is an interesting example of the privacy issues that wearable technology creates. Designed to work across wearable devices and phones from different manufacturers, Strava allows people to log their fitness activity – particularly cycling and running – using GPS. Users can create public profiles to share their activity with others and even engage in competition, such as setting records for fastest cycle ride through a given segment of road. Strava in turn anonymises and repackages this data for sale to a range of different commercial users. Local authorities, for example, can find out how many Strava users pass through a given section of the city, including anonymised age and gender data. A number of cities have used this information to inform decisions on transport planning (Walker 2016). Some researchers have looked at Strava data to examine the everyday practices of cyclists as a large cohort. Sun et al. (2017), for example, have demonstrated that people are more likely to engage in recreational cycling on residential areas and streets with limited traffic. A small study of this kind can be somewhat unfairly characterised as a very high-tech way to tell us what we already know, but it raises a more significant point as well; simply knowing that recreational riders stick to quieter streets does not tell us a great deal about cycling as an *embodied* practice. What is it about the *experience* of cycling that causes individuals to prefer some routes over others? Once again, the complex lives of individuals and the choices they make are reduced to simplified datapoints.

Strava places the responsibility on its users to manage their data safely. Thus, one can set up 'privacy zones' so that others cannot look at your profile and work out, for example, where you live and work. Some very good reasons for setting privacy zones became clear after a spate of bicycle thefts around 2015. Given that users could post the type of bike they ride on their public Strava page, it was a relatively easy matter for criminals to scan profiles looking for expensive equipment and not only work out users' addresses but in some cases even the garage the bike was normally stored in (Windsor 2015). Strava's default setting is for data to be open, making users vulnerable if they do not, *themselves*, think through the personal safety implications of not keeping their personal data private.

Some scholars have given considerable thought to how representations of these kinds of data can be built to protect users. A Finnish team has designed a web interface whereby a representation of cycling activity in cities such as Helsinki can be visualised in real time, with heatmaps showing the most popular routes at different times of day. They took care to exclude from their visualisation routes used by fewer than five people to mitigate the risk of revealing the homes of individual users (Sainio et al. 2015). This careful approach contrasted somewhat with Strava itself, which publicly released a global heatmap in 2017 that showed aggregated data from all its users. Unfortunately, quite a lot of military personnel use the app as part of their fitness training and thus secret military bases in remote parts of the world, including Afghanistan, showed up as bright spots in otherwise 'empty' countries. When Strava were criticised for this, their response was that it was the fault of the military personnel for not making their profiles private, rather than it being the company's responsibility to think through the potential consequences of what amounted to a massive uncontrolled release of the data it had gathered from the bodies of its users (Hern 2018).

As a platform, Strava is dominated by white middle-class men living in the developed West. Cycling, particularly within cities, is also dominated by the same cohort and as such Strava data is not entirely unrepresentative of a general population of cyclists. As a replacement for basic traffic counts, therefore, it is doubtless a useful tool for city planners. What is missing, of course, is an understanding of what would be needed to encourage more people, from a wider range of demographic groups, to get involved in cycling. Relying on this kind of dataset risks designing infrastructure around the needs of those who already cycle rather than those who do not – a reinforcement of existing privilege within urban space. Platforms of this kind reflect these wider issues of white male privilege, where the consequences of having one's privacy compromised by these kinds of monitoring technologies tend to be less of a concern. Fear of sexual assault, of being tracked down by a former partner, of being stalked and so on generally do not feature in the lives of people, frankly, like myself. For many it may not even occur that trading privacy for convenience can come with strings attached – if not for yourself then for other users. As researchers, therefore, we need to be quite cautious about the kinds of conclusions we can draw from datasets which have been constructed by the bodies of a specific group of individuals in a somewhat problematic manner.

Data transparency

The panopticon was a one-way system. The whole point was that the guards were hidden, so the prisoners never knew if they were being watched. Many of the technologies that threaten our privacy are similarly one-way – we become more visible, but the way information about us is used remains opaque. Cheerleaders for new technologies tend to emphasise the utopian qualities of greater transparency, without reflecting on the implicit power relations in this. There is a strong element of responsibilisation, where we are told that the production of more information about us allows us to make better choices about our lives. If you have better data about your patterns of exercise and sleeping derived from a wearable device, goes the logic, you will be empowered to live a healthier life. If city managers know which households in which neighbourhoods are producing the most rubbish because of monitoring devices placed in bins, citizens will be more likely to recycle and consume more carefully.

This kind of discourse draws on notions of 'nudge' architecture – incentivising individuals to make good choices rather than necessarily penalising them for bad ones (Thaler and Sunstein 2008). Like the panopticon, transparency here is one-way. Collecting information about the amount of waste produced by a household is intended to empower the city manager, not to provide citizens with evidence to argue for increased investment for services in their neigh-bourhoods. The utopian counterargument to this would be that we simply need to make this kind of data publicly available so that citizens can draw back the veil and challenge decision-makers. Indeed, this kind of logic has driven a really significant change in culture among a number of governments across the globe, increasingly operating on a principle of making government data open to all wherever possible.

On many levels this is a genuinely good thing and it also offers many interesting opportunities for researchers. There are, however, significant issues here in relation to power over who can *in practice* access and interpret these 'open' data sources:

> the public gains more information but, as a result, so do special interest groups. Transparency allows special interest groups to act quickly and influence decisions—actions that often bring about unfair outcomes to weaker population segments.
>
> (Zarsky 2016, 125)

This relates to similar issues with the free and open source software movement, which has significant methodological implications for critical scholars. With a software package such as R, for example, one can undertake tremendously powerful statistical and spatial analysis on enor-mous, public datasets that many governments now make available. Pro-grams like R are free to download for anyone to use and often require much lower powered (and therefore cheaper) computers than expensive proprietary systems. In theory, then, this allows the public to make use of the vast quantities of publicly available data now being put online.

Just because a piece of software does not require any financial investment to download it does not mean that it is genuinely *free*, however. R requires a tremendous investment of time and intellectual energy just to get to grips with its interface. It requires a fairly high-level understanding of mathematical concepts to even know what questions to ask with the software and, frankly, without significant external help, many people would really struggle to achieve anything meaningful with it. Yes, there are online guides available, but the barriers to use are high for all but the very geekiest – something that those of us working in research often forget. The existence of such open source software does not, therefore, neatly translate into empowering citizens to make use of newly available open data. As Zarsky indicates above, there-fore, openness and transparency in data can effectively reinforce existing inequality by privileging groups that already have significant expertise – or who have enough money to hire people with that expertise.

The scale of open data now being made available has created new dimen-sions to concerns around who gets to participate in a democracy underpinned by data, although it is a question that critical scholars have been grappling with for some time. Geographers working in participatory GIS (PGIS) have spent more than two decades trying to find ways of helping community groups engage with spatial data, creating maps to lobby local governments and others to provide more appropriate services in their neighbourhoods (Elwood 2006). I will discuss approaches to mapping in more detail in Chapter 6, but just one example here of a participatory planning experiment serves to illustrate some of the problems of assuming that having access to technologies and open data can meaningfully empower citizens.

Gordon and Koo (2008) asked participants to engage with the then popular 3D world making platform Second Life, getting them to envisage and virtually build an alternative Boston. As an exercise in getting participants to think through the challenges of remaking a city, the project was quite interesting. One of the significant things that it highlighted, however, was that the researchers spent a great deal more time than they expected simply getting their participants up to speed with how to navigate the 3D environment of Second Life, let alone actually building new elements within that environment. They suggested that follow-up projects would need to provide much more support to participants to help them engage with the platform so that they could begin to work with the virtual world being created as a tool to lobby for changes in the material world. Education and skills building thus became central even to engaging with something as relatively user-friendly as Second Life. In this light, telling non-experts that they can freely model, say, UK Land Registry data using R seems at best naïve.

Again, just to re-emphasise the point, there are tremendous power imbalances when considering nominally 'open' and 'transparent' data and platforms. As Muki Haklay (2013) has commented about the broader 'neo-geography' movement, there is an underlying utopian assumption that the technologies now becoming available can be taken advantage of by all, rather than sitting within a wider social system and thus reflecting existing exclusions and power relations. If you do not have the skills and access to the technology, data that is useful to more powerful groups remains entirely opaque to you.

This question of data transparency of course opens up lots of interesting methodological possibilities for researchers, particularly when working with community and third sector groups. Indeed, even relatively large organisations often lack capacity to analyse all the data that they gather. Dorothea Klein's lab at the University of Sheffield, for example, has collaborated with global charity Oxfam to re-analyse the treasure trove of survey and interview data gathered by their fieldworkers to shed new light on a range of questions of interest to the organisation (Tomkys Valteri 2018). As researchers, there are also interesting topics to explore simply in the ways that different organisations manage and analyse the data that they gather and what this does to their wider institutional mission.

Transparency and the algorithm

Data in and of itself, therefore, is of limited value. Data becomes interesting as it is analysed to show us new patterns and trends. The essence of so-called 'big data' is not really the size of the datasets drawn upon, but how algorithms have been produced to allow researchers to analyse and thus gain insights from these data (Yeung 2018). Scholars such as Rob Kitchin (2014) have for many years now been highlighting questions of accountability when decision-making is handed over to mathematical models. Algorithmic decision-making is an *actuarial* approach, where different variables are granted different weight within a model. Conventional human judgement, conversely, is characterised as *clinical* decision-making, where experience and intuition play a much greater role (Brauneis and Goodman 2018).

One of the reasons for the popularity of algorithmic decision-making is precisely because it appears to take out the element of human judgement, which is by definition subjective and open to bias. It is therefore seen as inherently more reliable. Of course, the problem with this is that the data which the algorithms examine are collected in a social context riven with bias – a point succinctly made by Lisa Gitelman (2013) in her book *'Raw Data' is an oxymoron*. More than this, however, the weightings of different variables in an algorithm are determined by human judgement but, with very few exceptions, the *composition* of algorithms being used in decision-making are opaque. In the facial recognition example discussed above, for example, we have no way to know even something as basic as the confidence threshold used by the system employed by South Wales Police to flag a potential match. Does the system need to be 100% confident of a match – thus increasing the risk that a potential offender goes unidentified – or is a lower threshold used, thereby increasing the risk of a false positive? We simply do not know.

Private companies argue, with some justification, that their algorithms have to remain confidential to prevent rival firms from gaining a commercial advantage. The difficulty with this, however, is that both as researchers and citizens, we have real difficulty in understanding how algorithmic decisions are made. In the Twitter example above, the company does not give details of how the 1% free dataset is sampled, so as researchers we have no way of knowing how representative it is for projects we undertake using those data. Even Facebook's own researchers concede that the way that the newsfeed is algorithmically created reinforces the 'filter bubble' effect that we generate by being friends with individuals similar to ourselves and liking stories close to our own interests (Lazer 2015). Independent researchers, however, cannot examine in detail how the algorithm reinforces the filter bubble, nor other effects, as the algorithms remain in a black box of commercial confidentiality. Striphas (2015) even argues that the rise of decision-making driven by algorithm fundamentally threatens culture as a publicly constructed pillar of our lives, instead constructing cultures via an opaque elite which we have no capacity to critically interrogate.

This is concerning because of the extent to which algorithms are presented as being neutral actors that give a truthful representation of the world free of human bias. Beer (2017) talks about this in terms not only of the code that was written to create the algorithm but also the extent to which the *idea* of the algorithm as a neutral actor has considerable social power. This can have particularly troubling implications as the neoliberal drive to reduce public spending has led to surprisingly large parts of the public sector being handed over to private companies working with data to inform local decision-making. For the most part the algorithms used are completely beyond public and democratic scrutiny, hiding behind claims of the need to preserve commercial confidentiality. They are presented as neutral, efficient and more truthful than a system based on the judgement of people.

Perhaps the most frightening example of this is within law enforcement. A number of firms now sell packages to police and judicial systems across the globe to assist in a range of issues from determining optimum patrol routes to judging the probability of recidivism by a newly released offender. Such algorithms are presented as being less subjective, but the danger is that:

> Minority neighborhoods historically subject to more intensive policing will have higher arrest and re-arrest rates, and then be recommended by the algorithm for more policing …
>
> (Brauneis and Goodman 2018, 125)

In a fascinating project using freedom of information requests, Brauneis and Goodman demonstrated that the US local governments they sampled had little or no understanding of how the algorithms ranked different variables or the thresholds for making recommendations. For the most part this was a product of commercial confidentiality clauses inserted by the vendors in contracts with local government, but also a simple lack of expertise and curiosity by those purchasing the systems.

As critical scholars there are a number of methodological responses to the lack of transparency around algorithms being used to inform very large elements of everyday society. Rob Kitchin's lab have undertaken a project where his team became deeply involved in the design and operation of a smart city data dashboard for Dublin and attempted to insert more critical perspectives into the design and operation of the underpinning algorithms and data management (Kitchin and McArdle 2018). Of course, not all scholars can operate at this scale and as Kitchin freely acknowledges, there are inevitable compromises to be made when working with very large commercial and public interests. As with Brauneis and Goodman, there are interesting projects to be undertaken simply exploring the extent to which those involved with algorithmic decision-making really understand what is happening 'under the hood' of the software they are engaging with.

From a playful methods perspective, there is great potential to engage with and attempt to subvert the algorithmic. This kind of approach can, of course, suffer the same criticisms directed at psychogeography (Debord 2006 [1958]), that it simply brings greater awareness of a social injustice to those who are already sympathetic, rather than really challenging the structures that have created that injustice. More playful approaches in this field tend to be underpinned by a digital ethnography, using the researcher's own experiences of an algorithmically constructed space in order to understand how those algorithms shape everyday understandings. Bucher's (2012) pioneering work on Facebook's EdgeRank algorithm is instructive here, examining how different materials are prioritised within the user's newsfeed. Indeed, Bucher's analysis throws up some really interesting insights into the extent to which the EdgeRank algorithm reworks notions of private versus public by positioning visibility as a goal to be aspired to by users – gamifying a regime of likes and interactions so that you 'win' by appearing at the top of your friends' news feeds.

It should also be borne in mind that algorithms are not stable entities, with platform designers reworking these to reflect and reshape user engagement. Wang's (2018) autoethnographic analysis of Chinese gay dating app Blued explores how users respond to and attempt to subvert the designers' intent. Most notably this was seen in attempts to game users' *yanzhi* scores (a measure of attractiveness) to increase the likelihood of a potential date getting in touch. This could occur through carefully curating the photographs uploaded to maximise qualities ranked by the algorithms as being of value (e.g. distance between eyes) or by altering the profile text to highlight qualities that do not feature in the app's predetermined categories of personality type (which excluded 'femme' qualities for example). Wang's explorations also highlighted how the designers had responded to the way users employed *yahzhi* within the app. Newer features being introduced explicitly connected to *yanzhi* as a means of giving users capital within the app, excluding men judged to be less attractive from the *nanshen* (male gods) streaming service. Thus, Wang's project shed light on the ways in which the 'neutral' algorithm was used as a tool for enforcing the app designers' understandings of what an attractive person was and how users should value themselves.

The Internet of Things

An Internet connected fridge that can detect when you are running low on groceries and automatically place an order with the supermarket. Lights that detect when no one is in a room and turn themselves off to save power. A speaker that plays the music you want just by talking to it. Various companies have pledged to deliver devices into your home that work 'like magic' to make your life simpler and more convenient. These devices have been broadly categorised as belonging to the 'Internet of Things' (IoT).

There are considerable privacy and safety concerns with IoT devices. Anything that is permanently connected to the Internet is a target for unscrupulous hackers – both by criminals and as part of state-led cyber warfare. There is no financial incentive for the manufacturer of a smart lightbulb or security camera to issue security patches for newly discovered vulnerabilities, meaning that users are installing highly compromised devices into their homes. In the UK there has been a long-running scandal in relation to the installation of so-called 'smart meters' for billing electricity and gas customers. The rollout of the second generation of these meters was repeatedly delayed amid concerns that they could be hacked to reveal customer details or even to allow a hostile state to cut off electricity supplies to households (Page 2018). This despite a multi-billion pound government backed initiative to create common and secure standards for these meters involving the National Cyber Security Centre, part of the government's spy agency GCHQ (Levy 2016).

If the might of GCHQ is struggling to secure an IoT device, then it is pretty much guaranteed that an inexpensive wi-fi enabled home security camera can be easily hacked. By installing such a device, you give a window into your living room for any criminal or government who is sufficiently interested to have a look. Again, having someone able to hack your smart lighting system may seem relatively trivial to some, but it is not hard to envisage a stalker or embittered ex-partner being able to engage in some modern day gaslighting by being able to remotely access and control devices in your home. Indeed, in 2018 a team at UCL put together a resource pack for victims of technology-facilitated domestic abuse precisely because of these kinds of threats (Tanczer et al. 2018).

From a researcher point of view, IoT offers some very exciting opportunities to engage in large-scale data collection about people's domestic habits. At present, IoT is a mess of conflicting standards, competing firms and a great deal of uncertainty. Where larger firms are involved – such as with attempts to create smart speakers and televisions that monitor their owners – it is very unlikely that the data generated will ever make its way into the hands of independent researchers. There are, however, very legitimate concerns with the ways that these devices gather and transmit data. Not only do many smart TVs report viewing patterns back to their manufacturers (and in some cases record audio and video of your living room), they often do so in an incredibly insecure way, which can easily be hacked by criminals and spy agencies (Vaughan-Nichols 2017). Being able to access your viewing habits is not only an invasion of privacy, allowing companies to classify the kind of consumer you are and sell that information to advertisers, it can also give insights to criminals about, for example, the best time of day to burgle your home.

Nonetheless, it is possible to take the IoT approach but use it to work *with* participants to explore their everyday domestic practices. An interesting example of this is some of the projects undertaken by Chris Speed's design informatics lab which combines an analysis of the outputs from IoT devices alongside ethnography with their users (for example, Cila et al. 2015, Vaniea et al. 2017). Such work can give participants insights into the kinds of data that devices gather about them and the inferences that can be drawn from them. In a small way, this allows the researcher to *give back* to their participants by increasing awareness of the

Figure 2.1 The Bitbarista accepts payments in Bitcoin for cups of coffee. The machine
can also offer to pay users in exchange for supplying it with services such as
cleaning the drip tray or adding coffee beans to the hopper. The project to
design and build the device was undertaken by Larissa Pschetz, Ella Tallyn,
Rory Giann and Chris Speed.
Source: Mark Kobine.

privacy implications of everyday technologies even while using those devices
to gain insights into their lives. More playfully, Speed's team also constructed
the Bitbarista, an Internet enabled coffee machine (Figure 2.1) which was able
to pay for new supplies using the Bitcoin users spent buying coffee from it,
opening debates about agency among connected objects (Pschetz et al. 2017).
As with many of the projects described in this book, Speed's approach requires
a fair degree of technical know-how in order to make it work. What projects
like the Bitbarista exemplify, however, is that there is tremendous capacity to
develop research methods that engage with new technologies while simultaneously challenging some of the more ethically problematic positivist
assumptions underlying their design.

Conclusion: the death of privacy?

When examining questions of privacy in our permanently connected and algor-ithmically monitored world it is easy to become somewhat despairing at the ways in which a dramatic erosion of civil liberties is being undertaken under the banner of convenience and safety. The project to roll out social credit scoring in a unified national system in China perhaps represents the most extreme example of this. Activities deemed not to be socially desirable by the state – for example too much time spent playing video games – could result in a lowered score, put-ting individuals at the back of the queue for healthcare, travel and a range of other services (Rollet 2018). This places a great deal of power in the hands of those who design the algorithm that judges what constitutes responsible beha-viour, with no transparency about how those decisions are made.

Reflecting on the issues raised in this chapter, there are four major oppor-tunities for research:

1   projects that involve collaborations with organisations harvesting perso-nal data, accepting some of the ethical compromises that come with this;
2   projects that explore the impacts of privacy-threatening technologies on everyday behaviours;
3   projects that seek to raise awareness of these threats to privacy, educating users in how to protect themselves and lobbying for change;
4   projects that engage with these technologies to devise novel research approaches that challenge or subvert some of the assumptions that underlie them.

Very interesting work is happening in the first two categories and it is by no means my intention to dismiss these kinds of projects. Nonetheless, the underlying approach of playful methods means that this book primarily con-centrates on the third and fourth areas. As such, playful methods has a somewhat interventionist character, not simply trying to develop new research techniques, but actively attempting to do so in a way that resists neoliberal attempts to control and exploit citizens for financial gain.

Privacy and transparency are underlying themes running throughout this book. As a point to end on here, I want to come back to a consideration of the individual *bodies* that produce and, increasingly, are produced by the data being gathered about us all. As Cheney-Lippold (2011, 170) notes:

> algorithms allow a shift to a more flexible and functional definition of the category, one that de-essentializes gender from its corporeal and societal forms and determinations while it also re-essentializes gender as a statis-tically-related, largely market research-driven category.

When Facebook classifies a user as 'female' that information is useful not in terms of the *experience* of being a woman in the world today, but as a tool for determining what kinds of things they are likely to buy – reducing the body to a mere vehicle for consumption.

The danger, then, is of losing sight of the *embodied* when we develop research techniques based on these new technologies. The drive to undermine privacy is an excellent strategy for selling products, for cutting costs in local government, for making life more convenient if you are privileged and for enhancing the power of elites. This makes engaging with these technologies difficult for critical scholars. It becomes incredibly challenging even to secure meaningful consent from participants as many people have no idea just how much can be understood about them as different technologies capture traces of their bodies. Using data gathered at scale in conjunction with algorithmic analysis allows us to tell interesting stories about wider society, but in treating individuals as data points we occlude the richness of life as it is lived. Ultimately, therefore, we have to acknowledge that these approaches are simultaneously distanced from the body and yet intimate. Responding to this tension is at the heart of the methodological approaches that are discussed in this book.

Note

1   Martin Hawksley's TAGS project: https://tags.hawksey.info/.

3 Measuring the body

Phil Jones and Tess Osborne

Introduction

We start this chapter rather awkwardly with a note about its authorship because, in contrast to the rest of the book, this part is co-written. Writing this as a collaborative chapter acknowledges that Phil only became interested in questions around measuring embodiment as a direct result of working with Tess. Much of the section below on advanced physiological measurement draws directly on Tess' PhD research, some of which we have jointly published, with other material authored by Tess alone. Working with Tess undoubtedly drew Phil to become interested in an area of research he had not previously considered. This in itself is important as it serves as a reminder of how collaborative work can stimulate new ideas and take research in novel directions.

The previous chapter highlighted some of the privacy concerns raised by the use of wearable technologies. Here we consider what such devices might mean for the practice of research, particularly in the way that they can be used to quantify different elements of our bodies. Of course, researchers have been measuring characteristics of the body for centuries, but the types of tools at our disposal have increased significantly in recent years. Where high quality physiological measurement was previously only readily available to medics and those working in psychology, new technologies have given consumers easier access to these indicators. This provides critical researchers with new opportunities both to study the ways in which these technologies are being used and to deploy them within our own research projects.

The step counter is one of the simplest devices available today to measure aspects of our bodily activity. There is something slightly hypnotic about watching the numbers tick upwards as you go through the day. These devices can generate a huge sense of achievement when the daily target is reached and reinforce a sense of failure when it isn't. Beyond simple step measurement, it is becoming increasingly straightforward to measure heart rate, blood pressure, skin conductance and temperature. Commercially available kits allow people to examine their cortisol levels as a measure of stress or to calculate their ethnic makeup via DNA testing. There are devices available that measure brainwave activity and eye movement, which, although still expensive, have fallen sufficiently in price to be realistic options for use by non-specialist researchers.

All these new technologies offer routes into potentially interesting research projects. A key theme which recurs throughout this chapter, however, is that while it is becoming ever easier to gather *data* about embodiment, it remains a challenge to *analyse* those data in a meaningful way. While there are tools available to help interpret these measures – allowing us to infer emotional response for example – they need to be used with a great deal of caution. The body cannot be seen in isolation and it is crucial to understand the context in which measures are being taken. This is true both at the small scale of examining particular stimuli and at the larger scale of understanding the wider socioeconomic structures and injustices that shape everyday life.

The chapter opens with a consideration of the quantified self and the desire for self-knowledge through measurement. This has been driven by a market selling ideals of fitness and well-being, usually to broadly healthy people who actually need this kind of digital assistance least. We go on to consider biohacking as a more extreme form of the desire to assess and adapt the body, optimising its performance. There are some clear gendered considerations here and deep ethical concerns that would keep critical researchers away from biohacking as an approach to research rather than simply a topic for examination.

The second half of the chapter turns to examine different kinds of physiological measures, including eye tracking, to reflect on how critical researchers can employ these within projects. Again, we emphasise the importance of contextualising these measures in how we interpret the data that they produce. We also explore the limits of non-specialist application of these approaches and reflect on how experimentation with these techniques can lead to more productive discussions with experts to develop follow-up projects. Critical researchers are right to be cautious about the claims that are sometimes made for what these data can tell us about the individual. Nonetheless, these datasets can give surprising insights into people's responses to stimuli and there are significant issues to consider around the ethics of this kind of work, particularly questions of informed consent. This is a field that seems to attract a great many charlatans working outside the constraints of the university sector which puts an even greater onus on academic researchers to debunk these claims and undertake their own research in an ethically rigorous manner.

The quantified self

The quantified self is both an approach to self-monitoring and the name for a loose grouping of individuals who have championed this activity. Gary Wolf (2016) claims credit for co-organising the first 'quantified self' group meeting in 2008, held in Pacifica, California. Pacifica is a city neighbouring San Francisco and the connection to the Silicon Valley tech scene runs deep within ideas of quantifying embodiment. As Nafus (2016) points out, at the heart of the quantified self is an understanding of the individual as being ultimately responsible for taking control of their lives. This type of belief lends itself to neoliberal fantasy, ignoring how governments, corporations and others dictate the framework in which we live. If we are overweight, for example, this discourse would suggest that our fatness is the result of our own lack of self-control, rather than the fact that we live in an obesogenic society.

Although the phrase 'quantified self' is of relatively recent origin, the idea of self-monitoring has a somewhat longer history. Arguably different forms of personal archiving – keeping photo albums, retaining correspondence and so on – can be seen as practices of capturing and retaining manifestations of the personal and embodied. Even a simple weight-loss diary or marking a wall to show your children getting taller on each birthday reflect this desire to show changes to the body over time. In the early 1990s we saw attempts to use what were then novel digital technologies to explore different ways of capturing the bodily over time. MIT's Wearable Computing Project, established in 1992, set the template for this kind of work. This included the early use of digital cameras for 'lifelogging' whereby a large number of images were captured every day from a device worn on the body to create a kind of digital diary (Lupton 2016).

As discussed in the previous chapter, the consumer market for wearable technologies arose in part from the tech sector's attempts to find the next big thing. Wearables remain far from ubiquitous despite the hype and multi-billion-dollar marketing budgets. This is partly because, for the vast majority of people, wearables remain a solution in search of a problem. Nonetheless, if these devices have a useful function it is in taking real-time measurements of our bodily activity. The technology comes in a variety of forms, with market-ing teams adding the word 'smart' to watches, glasses, clothing and even implants. The vast majority of consumer wearables only monitor a fairly limited range of physiological measures: some measure heart rate; the Apple Watch can create electrocardiograms; some use GPS to record running/ cycling speed; and many simply measure the body's movements using gyro-scopes. As a general rule these do not require specific medical expertise to interpret although the gyroscope data is rarely made available to users in its raw form – we are therefore reliant on the manufacturer's commercially con-fidential algorithms to interpret what the recorded movement means.

Of course, in highly specialised applications, wearable smart devices can bring immense benefits. The Freestyle Libre, for example, uses a sensor implant with a probe that sits just below the surface of the user's skin (Figure 3.1). This allows diabetics to take regular blood sugar readings simply by placing a wire-less device near the sensor as an alternative to traditional pinprick tests. The system automatically creates graphs for patients and their doctors to monitor fluctuations blood sugar levels over time, helping them to put in place more effective control strategies (Fokkert et al. 2017).

Devices like the Freestyle Libre can have a genuinely transformative effect on life for people with specific medical conditions. This has driven a degree of

Figure 3.1 The Freestyle Libre employs a replaceable sensor using near field commu-
          nication (NFC) to transfer blood sugar readings to a custom device or
          mobile phone app.
Model: Morag Robertson.
Source: Author.

excitement about the possibility of routinised personal monitoring as a means
of driving preventative approaches to health. Swan (2012, 93), for example,
has described this rather grandly as the rise of the 'participatory biocitizen'.
Medical grade monitoring remains somewhat niche, however, and the use of
consumer-grade wearables has been driven almost exclusively by the fitness
and 'wellness' markets. Devices like the Apple Watch and the Fitbit have
brought ideas of quantifying bodily response into greater public consciousness. The discourse that has arisen around these wearables is that they enable
users to take more control over their bodies by quantifying their daily activities, allowing them to uncover things about themselves, then set and achieve
goals for self-improvement (Lupton 2016).

Research by Public Health England (2018) indicates that 22% of all adults
in England can be considered physically inactive. This figure is significantly
driven by social class, with only 18% of the richest quintile being inactive
compared to 26% among the poorest. Of English adults aged 40–60 (over 6m
people) 40% do not even manage a 'brisk' walk for ten minutes per month.
Some researchers have explored the possibility of using fitness trackers in

combination with incentive schemes to try to tackle the health challenges posed by an increasingly inactive society (Shin et al. 2016). For the most part, however, the kinds of people who buy and regularly use these fitness devices are not those who would be the public health priority for increasing physical activity, with users more likely to already be health conscious (Shin and Biocca 2017).

The phenomenon of the 'worried well' has long been recognised within the medical profession, with very minor symptoms being believed by some patients to signal a much more serious illness (Wagner and Curran 1984). In many ways the fitness tracker panders to the worried well, giving users the opportunity to fret when they miss their daily target for steps taken, calories burned or hours of deep sleep. Indeed, sleep tracking has been demonstrated to be particularly problematic in this regard, not only because consumer devices are notoriously inaccurate (Kolla et al. 2016) but also because people worrying about their trackers reporting that they have not had enough sleep has been shown to exacerbate *actual* sleep problems (Baron et al. 2017).

Control is key here. Those who self-monitor (including the extreme 'biohackers' discussed below) are seeking control over themselves. Conversely, some businesses have started to use this kind of monitoring as way of controlling their employees. Of course, there can be benefits for employees using wearable technologies: warning construction workers when they enter a dangerous area; alerting a driver if they are starting to lose concentration; helping staff to maintain accurate posture to prevent long-term back problems and so on (Khakurel et al. 2018). The temptation for employers, however, is to use these devices in a more punitive manner. Many large warehouse operations, for example, now routinely monitor their staff through wearables to ensure that they hit targets for picking, with breaks and decreased productivity rigorously policed under threat of dismissal (Moore and Robinson 2015). Indeed, the technologies associated with the quantified self are now routinely identified with the precarity that characterises the neoliberal workforce (Moore 2017).

A useful approach to research in this kind of area is in getting participants to talk through the data that the devices generate about them. For those whose employers require them to be monitored, this is an opportunity to regain some small measure of control over that data, although such a project would require the collaboration of the employer which may not be possible when undertaking explicitly critical research. For those who self-monitor, it is easier to access their bodily data, although there are some ethical considerations that need to be taken account of as the researcher gains access to what in some cases is a form of medical data. This is particularly important where projects are taking a more interventionist approach, for example, temporarily loaning participants a device that they would not otherwise use. Here, giving participants the opportunity to see the data they have generated is crucial, but it also raises potential problems where the researcher is not a medical expert. Researchers need to manage participant expectations and explain that they are not qualified to interpret or diagnose any potential health risks that are shown by the data.

Researchers also need to consider the assumptions that underpin the ways in which the quantification of bodies is enacted. Lupton (2015), for example, has looked at the ways that apps designed to quantify sexual and reproductive health make unquestioning assumptions of heteronormativity. Such apps promise much greater accuracy than traditional practices of menstrual tracking but some of them also add rather cringeworthy elements, such as sending messages reminding men to bring home flowers when their partner enters her fertile window. Apps targeted at men tend toward gamifying sex – duration, number of partners etc. – where those targeted at women emphasise risk and health. The fact that problematically gendered assumptions are 'baked in' to the apps and devices that drive the quantification of bodies needs to be actively reflected on when we undertake research employing these technologies.

Biohacking

Biohacking has been defined as 'the practice of manipulating biology through engaging biomolecular, medical, and technological innovations' (Malatino 2017, 179). Biohackers employ a range of devices and chemicals to monitor and augment their bodies. Treating the body as if were computer code that can be tweaked and optimised comes with considerable issues around ethics and social justice, however, because only the relatively wealthy have the time and financial resources necessary to explore these possibilities. The different types of biohacking are united by the idea that the body can be measured, reworked and customised to escape the constraints of nature. It is interesting to reflect on these discourses through the lens of Donna Haraway's (1985) cyborg manifesto. Haraway saw a blurring of humans, animals and machines through two lenses: optimistically, that this process might create more empathy and kinship with the non-human world; and more pessimistically, that this process reinforced violent masculinity and, implicitly, neoliberal privilege.

Some forms of biohacking appear to be a not much more than a macho way of describing healthy living – taking appropriate exercise and consuming a balanced diet – even if different technologies are employed to assist with this process. There is, however, a significant problem of modern snake oil salesmen promoting diet plans, supplements and drugs such as anabolic steroids as part of biohacking regimes that are at best useless and at worst actively harmful. A recent article in *The Lancet* has drawn parallels between extreme approaches to biohacking around diet and muscle building in young men and anorexic-type eating disorders in young women (Nagata et al. 2019). Silicon Valley's embrace of the Paleo diet as part of a technofetishist discourse of hacking is merely the latest manifestation in a long line of fad diets where people have sought to gain control over their unruly bodies (Gioia 2015).

Biohacking is dominated by men and has more than a whiff of the Nietzschean *Übermensch* fantasy about it (Lukyanov 2019). Individuals like Serge Faguet grab the headlines – a Silicon Valley entrepreneur who mixes different technological augmentations to his body with a variety of 'anti-aging' drugs because he intends to live forever (Marsh 2018). The 'tech-bro' connection can also be strongly seen in phenomena such as the garage biology movement, which plays on the Silicon Valley origin myth of the garage-based start-up business that goes on to change the world. Garage biology takes the life sciences out of its conventional lab settings, with a mythos of rebel scientists taking on the biosciences establishment (Delfanti 2012). Some of this work can be quite interesting, for example creating kits to crowdsource analysis of bacterial samples gathered from the buttons of pedestrian crossings in different cities. This is intended to help researchers analyse the temporal and spatial distribution of microbial life to answer questions around disease vectors and public health (www.bioweathermap.org). Other elements of garage biology are more problematic and at times downright dangerous, because they are predicted on the idea of disruption (Nash 2010). The disruption model of 'break things and apologise later' is unrepentantly neoliberal. It is one thing for a company like Uber to drive local taxi firms out of business while running massive losses propped up by venture capital. It is quite another thing to 'disrupt' the safety and ethical protocols that are required in formal biosciences laboratories by claiming that they are a barrier to innovation.

The garage biology mentality has driven the rise of businesses encouraging people to take tissue or fluid samples themselves either for self-testing at home or to send off for lab-based analysis. Some of this is about cost saving and convenience in routine medical screening, for example, self-sampling for HPV, though this approach is regarded as much less effective than going for regular pap smears (Farnsworth 2016). DNA testing for ancestry or predisposition to certain genetic diseases has also become fairly common. Knowing you have a predisposition for a particular disease is not necessarily terribly useful information to have and it can be quite upsetting to find out about this without the contextualisation that might be offered by a medical professional. Ancestry testing, meanwhile, hit the headlines in 2018 when Senator Elizabeth Warren, goaded by President Trump, released results of a DNA test that demonstrated she could trace a very small branch of her family tree back to a native American ancestor (Lartey 2018). Some firms involved in this DNA testing use the samples that they gather as part of a process of building up a biobank, which allows researchers to explore genetic processes across a wide sample of the population. In essence, by seeking to quantify elements of their DNA, users of these testing kits are unwittingly undertaking biopolitical labour, providing biomedical businesses with commercially valuable data about their bodies (Duster 2016).

An alternate reading of garage biology might be to examine the ways in which it has been used to support marginalised groups. The prime example here is within the trans community. For some trans people, transition comes through wearing their bodies differently, changing clothes, names and so on; others pursue a biochemical route. A garage biology approach can, for example, enable a trans person to bypass the formal strictures of the medical profession in order to remake their body through privately sourced supplies of hormones (Malatino 2017). Nonetheless, other forms of personal biochemical alteration are arguably more 'lifestyle' orientated, such as the use of so-called smart drugs. Modafinil, for example, was designed to keep soldiers awake on 48-hour missions but has subsequently been marketed to those working in high pressure corporate jobs with tight deadlines (Sharma 2014). Adderall, meanwhile, is used to treat attention deficit disorder in children but has become popular among students as a means of enhancing concentration (DeSantis et al. 2008). Universities and colleges have seen a rise in the use of smart or study drugs in the last two decades although a recent meta-review found that the use of different stimulants was probably more effective in driving the motivation of individuals than actually expanding their cognitive ability (Ilieva et al. 2015). This, of course, raises ethical questions for educators worried about off-prescription drug use and the pressure that their widespread availability on campuses puts on students desperate to succeed.

Other forms of biohacking involve the use of implants. In a medical context these might include something like a pacemaker or an insulin pump to compensate for a chronic health condition. Some biohackers (so-called 'grinders') use implants instead as a means to augment their bodies. An example of this is surgically embedding subcutaneous magnets which allow users to pick up ferrous objects without gripping them. On the one hand this is a kind of body modification analogous to piercings and tattoos, but the magnets move when exposed to an electro-magnetic field, creating a fizzing sensation in nearby nerves which some grinders describe as being like an extra sense (Doerksen 2017). Perhaps more interesting from the perspective of this book is the use of radio-frequency identification (RFID) tags. RFID is the technology underpinning contactless credit card payments, comprising a small, passive chip that can securely transmit information over a short distance. The first implantable RFID tag for use in humans was approved in the US in 2004 (Rotter et al. 2008). Some businesses have subsequently supplied employees with subcutaneous RFID chips so that they can access different buildings and systems by simply waving their hand over a sensor. This has significant security advantages over conventional ID badges because the implanted chip cannot be easily lost or stolen (Michael and Michael 2012). Some enthusiasts have chosen to deploy these systems for personal use such as securing their own homes. These systems make it easier to innumerate the flow of people in and out of different areas, which can be particularly useful in high-risk industrial environments, enhancing employee safety. Of course, they also bring the potential for more punitive surveillance by the employer.

One does not need, of course, to implant a piece of technology for it to augment the body, converting it into an object which can be more easily monitored and measured. The vast majority of people in developed countries now carry smartphones. These devices can be used by other people to monitor us without necessarily gaining our consent – arguably a form of biohacking that is controlled by outsiders. In a fascinating proof-of-concept study, an Australian team examined how university staff were using a break room by analysing mobile phone signals (Abedi et al. 2014). Most people leave their phone's Wi-Fi permanently turned on so that it seamlessly connects to known networks. Wi-Fi routers can detect the presence of a phone, even if it is not connected to that router, because the phone is constantly scanning for available Wi-Fi networks. When phones do this the Wi-Fi router can detect their phone's MAC address. Abedi et al. placed a router in a break room to create a log of every time an individual phone came within around 15 metres of it. As a result, they were able to detect how many people were using the break room at any given time, what the busy times of day where and how long individuals spent within it.

Systems to harvest users' MAC addresses were sold to shopping malls as means of tracking customer flows. This is, however, an unusual example of the tech sector stepping in to provide more security to their users. More recent iPhones randomise the MAC addresses that they broadcast when they are not actually connected to a Wi-Fi network. This has effectively broken shopping mall Wi-Fi tracking systems except for the small number of customers who actively connect to the mall's free Wi-Fi service (Prasad 2018).

There is a danger that whenever we use technology to enhance our bodies, those controlling the technologies gain information about us. Period tracking apps are sold as giving their users more control over their health and fertility. It is, however, depressingly common for the data stored in these apps to be sold to third parties (Burke 2018). This raises questions about whether this is really a case of the individual gaining control over their body or ceding control to corporations who can optimise the timing of advertising pushed to users' phones based on their patterns of menstruation. In short, technological augmentation does not necessarily always work on our behalf.

From a research point of view, actually employing biohacking practices within projects comes with major ethical concerns. Even the kind of non-invasive bodily tracking using mobile phones described by Abedi et al. needed careful consideration because of its potential for privacy violations. Biohackers themselves might argue that the formal strictures of social science are ripe for disruption and that we should be considering 'action research' using implants, chemical supplements and the like to modify and track our participants. University ethics committees would, of course, take a dim view of this kind of disruption. Thus, while biohacking as a *practice* is ripe for investigation through conventional social science techniques, biohacking as a *technique* for research, adapting the bodies of participants is clearly something that we could not endorse. Garage approaches do not sit well with the ethics of critical social science.

Advanced physiological measures

One of the truisms of social science research is that the material we gather from our participants is always situated and partial. People lie to researchers, they misrepresent their views, they tell us what we want to hear. Frequently they may not be able to articulate or rationalise why they have taken a particular decision or believe something to be the case. The views they give always come from a subject position and people may protect themselves in how they speak about particular issues. This happens regardless of the methodology we choose, from large quantitative surveys, interviews of different kinds, focus groups through to intensive participatory research; we have no magic route into participants' heads to capture their true feelings.

Of course, to make sweeping generalisations as we have in the previous paragraph is to ignore decades of social scientific work within psychology investigating physiological response. The body itself can give insights into a participant's thoughts and feelings and a range of techniques have thus been developed to measure different elements of our embodiment. Some of the techniques developed for psychological and medical applications have started to find their way into other areas. As discussed above, consumer-grade wearables can give interesting if limited insights, but there are now a variety of new devices that have started to take advanced physiological measurement out of lab settings and make them available at more accessible price points for non-clinical researchers (Neff and Nafus 2016).

Of course, some of these body-sensing technologies are still very much the sole purview of specialists: Functional Magnetic Resonance Imaging (fMRI), for example, remains extremely expensive and complex. It has been used to investigate a huge range of research questions by, very crudely, seeing which areas of the brain 'light up' in response to different stimuli. Such technology cannot 'read' our thoughts, however, and although it is undoubtedly underpins an exciting field of research there is an argument about the limits to the conclusions which can be drawn from such studies (Klein 2010). Collaborating with an expert would be the only viable way for most researchers to access the insights this approach can bring. Nonetheless, other advanced devices for measuring bodily activity are becoming ever more affordable to non-specialists.

An important caveat to set out at the start of this section is that although these devices are more widely available, they still require researchers to develop significant expertise to analyse and interpret the data that they produce. Nonetheless, one of the reasons why scholars beyond psychology are beginning to become interested in these devices is because they offer the potential to gain insights into participants' emotional state (Evans 2001, Picard 1997). The claim that we can capture emotional response doubtless would make many critical scholars uncomfortable and so it is important to insert some caveats here. Effectively these technologies only allow us to draw *inferences* about emotional state from the data rather than actually *measuring* emotions. Likewise, the data generated by these devices can only tell us that the body is responding in a particular way, they do not explain *why*.

The umbrella term 'biosensors' can be used to describe devices for measuring different qualities of the body. Broadly, these fall into two types, 'wet' biosensors which measure the presence of particular chemicals and 'dry' biosensors which non-invasively monitor different signals generated by the body. The Freestyle Libre described above would fall into the category of a wet biosensor because it has an implant measuring bodily chemistry. When seeking to examine emotional state, one of the most useful 'wet' measures is cortisol levels. This chemical is released in the body in response to stressful situations. It can be measured immediately through saliva but also leaves a long-term trace in hair follicles, such that variability of stress over a period of months can be examined for people with longer hair. There are some issues with using cortisol measures, however. They can, for example, be affected by previous exposure to trauma, with a study showing that women exposed to physical abuse as children have a significantly lower cortisol response to standard stress tests (Carpenter et al. 2011). Wearable patches for real-time, non-invasive cortisol measurement have been tested in the lab but are still some way from being commercially available (Parlak et al. 2018). As a result, cheek swabs or hair samples must be taken from participants and sent to a lab for analysis. This makes the use of cortisol measures significantly more complex to operationalise than 'dry' measures, particularly in terms of the practicalities and ethical implications of how biological material is gathered, stored and transferred.

For most scholars working outside a clinical context, 'dry' biosensors are the only real option. These require no more than contact with the skin and can capture a range of different bodily signals from skin temperature and heart rate to brain waves and electrical activity in the muscles. When taking these devices outside the controlled confines of a lab, however, there are so many different environmental factors at play that it can be very difficult to isolate which stimulus a participant's body is responding to. A nice example of this problem comes in an innovative project by Aspinall et al. (2015) which used a mobile electroencephalogram (EEG) device to examine the response of participants walking around central Edinburgh. The EEG measures brainwave activity through a series of sensors arranged around the skull of participants and in recent years mobile versions have been made available. Aspinall et al. used a version of the EMOTIV EPOC+, a device which the manufacturers advertise as generating 'research grade results' (Figure 3.2). Although the EPOC+ itself is not particularly expensive ($800 at the time of writing), the manufacturer has a subscription-only model for its advanced analytical software which adds a significant ongoing cost.

The different frequencies of brain wave activity captured by the device can be mapped against emotional states from deep sleep and autonomic tasks through to excitement and agitation. The EPOC+ uses a series of (commercially confidential) algorithms to classify emotional state based on the signals recorded by its sensors. Aspinall et al. wrote a custom piece of software to link the processed outputs from the headset to the participant's GPS location. The 12 participants in their project were asked to walk a fixed route around the city, taking in three different types of urban environments – busy shopping street, green space, noisy commercial district. The researchers

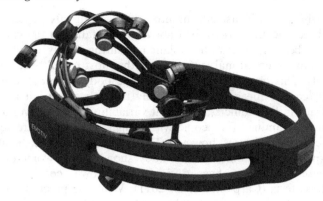

Figure 3.2 The EMOTIV EPOC+ is an off-the-shelf device for measuring EEG data
in research settings. Access to its higher functions requires subscription to
the company's advanced analysis software.
Source: EMOTIV.

were able to show a measurable difference in the emotional states generated
by the three environment types. In particular when participants entered the
green space zone, the researchers saw a reduction in brainwave activity associated with arousal and frustration as well as an increase in meditation.

This last point is interesting, as Aspinall et al. point out, because it is in
line with theories around the calming and restorative effects of exposure to
green space (Ulrich 1984, Kaplan 1995). This is, however, one of the only
explanations suggested by the research team for the observed brainwave
responses to the different field sites. We cannot extract from these data what
the effect of, for example, air quality, traffic, a love of Victorian architecture
or a range of other factors has had on the participants' physiological
response. There is simply too much going on in a complex urban environment
for us to isolate particular stimuli producing specific patterns of brainwave
activity. It is also important to note, as Aspinall et al. themselves do, that
sensors can be displaced by participant movement – as one would expect
when moving around normally in a non-lab environment. Displacement creates false signals which can be interpreted in unpredictable ways by the automated algorithm embedded in EMOTIV's analysis software. This serves as a
useful reminder that the data from these devices needs careful interpretation.
It is imperative, therefore, to avoid the temptation to treat these devices as
'black boxes', investing the data that they produce with the aura of unvarn-
ished scientific truth.

When employing wearable devices with participants, particularly in a field
setting, there are advantages to these being as discreet as possible. This pre-
vents participants being self-conscious about wearing complex equipment that
might draw comment from passers-by. Less invasive equipment can also help

participants to forget that they being monitored, as well as being easier to set up. Thus, there are compromises to be made. The Bioharness, for example, was developed to monitor high performance athletes and is strapped onto the user beneath their clothing. The device allows the measurement of breathing rate and by placing the heart rate sensor directly on the chest gives highly accurate results. The system wirelessly streams data to a remote monitor so that coaches can give real-time feedback on an athlete's performance. It can also be used with a disposable core temperature sensor which participants swallow in advance of the planned activity. While such a device gives highly accurate data, particularly in exercise scenarios (Johnstone et al. 2012), in common with the head-mounted EMOTIV EPOC+ participants cannot help but be aware of the presence of the device.

By contrast, one of the most popular devices to have reached the research market in recent years is the Empatica E4 (Figure 3.3). Worn on the wrist it has the appearance of an unremarkable sports watch and it is easy to forget that you are wearing it. The sensors give slightly less accurate data when readings are taken from the wrist than they would if mounted in optimal

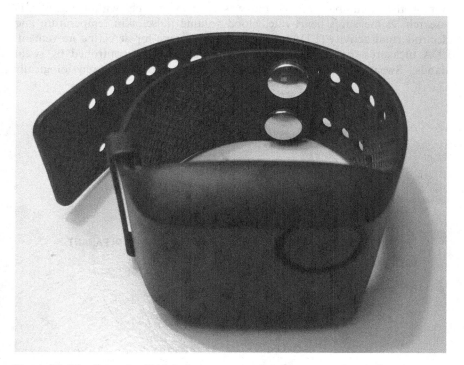

Figure 3.3 The Empatica E4 has become one of the most popular devices for accessing participants' physiological data, helped by its simplicity of operation and user-friendly analysis tools.

Source: Author.

positions (e.g. chest mounting for heart rate). Trading off a slight reduction in data quality for a discreet, well packaged and easy-to-use device is an acceptable compromise for all but the most specialised projects.

We have found that Posner et al.'s (2005) interpretation of Russell's (1980) classic circumplex model of affect is a valuable tool to use with students when exploring how to infer emotional state (Figure 3.4). Russell argued that different emotional states could be seen as being related to each other, arranging them around a circle to illustrate their connections. The layout of the circle was based on two axes, one signifying the degree of arousal and one signifying the degree of pleasure. Traditional cognitive theories of emotion argued that discreet neural systems deal with each emotion. In a review, however, Posner et al. found that the circumplex model was a better fit with the empirical evidence that has since been generated by neuroimaging and cognitive neuroscience. The upshot of this is that combining measures of physiological arousal with measures of pleasure gives a useful toolkit for inferring the broad types of emotions being experienced by a participant at a given moment.

The E4 lends itself to generating the arousal/pleasure measures needed to employ the circumplex model and was used extensively within Tess's PhD research. It measures heart rate, blood volume pulse, skin temperature and electrodermal activity (EDA) alongside a gyroscope for detecting movement. EDA measures the skin's electrical resistance which is controlled by sweat glands. Sweat gland activity increases when bodies are physiologically

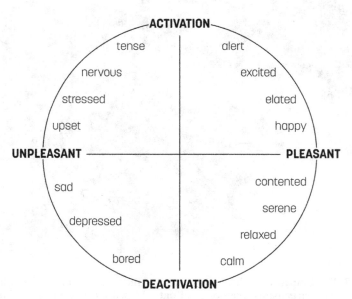

Figure 3.4 Posner et al.'s (2005) circumplex model of affect.
Source: Redrawn by Chantal Jackson.

aroused. Thus, EDA serves as a useful measure for the activation axis on the circumplex model. Similarly, skin temperature can be used to examine the pleasant/unpleasant axis, although this is a little more complex to interpret as a slight rise (flushing) can be associated with somewhat unpleasant feelings (Vos et al. 2012) but a large drop reflects strongly negative feelings (Herborn et al. 2015). Vos et al. also identified a sudden drop in heart rate as being associated with negative emotions in the six seconds following a particular prompt.

The idea of using physiological measurement as a tool for mapping urban space has been around for some time now. Artist-designer Christian Nold's 2004 Biomapping project used a homemade galvanic skin response sensor (a similar measure to EDA) in combination with a GPS to map arousal levels of participants walking around their neighbourhoods (Nold 2009). Crucially, however, he gave the participants their own data to examine – represented as a three-dimensional overlay in Google Earth – asking them to explain what was happening at different points on their walks. Some of what was revealed would have been easy to predict from the map – a peak of arousal when crossing a busy road for example. Other elements would have been impossible to work out without the participants' own insights, such as reacting to a group of noisy children. Environmental context, supplied by the participants, was therefore crucial to interpreting the data.

Nold's homemade wearable lacked a measure for the pleasure axis needed to employ the circumplex model, a limitation which has subsequently been overcome by devices such as the E4. We have written elsewhere about the potential for using this kind of device for inferring emotional state in field studies (Osborne and Jones 2017). Our project compared measurements taken from individuals walking around their local neighbourhood to participants undertaking a proxy walking exercise using a video game in a lab-type setting. The activity of both groups was filmed and follow-up interviews undertaken in order to contextualise the measures recorded via the E4 wristband. We found that only around a third of participants in both groups actually generated clear peaks and troughs in the signals recorded. Of the remainder, most saw only general trends of indicators rising and falling and some, particularly in the lab study, saw very little variation at all in their data. With some of the recordings made in both field and lab settings it was possible to pick out events from the video and the interviews that mapped onto patterns seen in the data from which emotional state could be inferred using the circumplex model. Examples of this could be seen in a participant in the field study recalling a dark time in their past while walking on a particular street. Similarly, there were clear excitement responses by participants in the videogame study when they accidentally found themselves caught in a conflict situation.

If it were possible to reliably capture emotion from these physiological measures, the ideal would be that a large number of participants could be asked to walk around a neighbourhood while wearing the device and from this a scientifically 'objective' map of emotions could be derived. There are, however, unsurmountable problems in making that ideal a reality. Partly this is because these sensors simply do not deliver clear and unambiguous signals which can be unproblematically converted into an interpretation of emotional state. Second, even if the reliability of the measurements could be improved, there is still the same problem that was seen in the Edinburgh example examining brainwave activity; it is simply not clear what elements in a complex real-world landscape generate a particular physiological response.

Bernd Resch's lab has run a project attempting to overcome some of these issues. The Urban Emotions project combined a dry biosensor with a smartphone app (Resch et al. 2015, Zeile et al. 2015). Participants were given the opportunity to 'ground-truth' spikes in their biosensor data (which measured heart rate, EDA and skin temperature) via a real-time alert on their phone. The app allowed participants to give a brief explanation of what might be causing the physiological response in that moment. For the sake of simplicity, the explanations were limited to a single word (e.g. 'traffic') in contrast to the rich interview narratives that we elicited on the project described above. Again, this is a question of compromise. To generate instant responses in a field setting, short answers are required; it simply would not be practical to ask participants to type a small essay into their phones while walking around an urban environment. Nonetheless this gives an opportunity to provide some immediate contextual data which, because of its relative simplicity, could then be subject to automated classification within the project analysis.

In a fascinating paper Shoval et al. (2018) asked 68 tourists to wear the E4 wristband while walking around Jerusalem on their first or second day in the city. The EDA data recorded were normalised for each participant and used to create a series of cells showing the level of arousal across the whole participant group in different parts of the city. The authors acknowledge that while there could be multiple explanations for why some sites might be more arousing than others, they noted that religious sites and areas with security risks seemed to have a particularly strong effect.

When we discussed this the project with the lead author at a conference, we asked whether there had been any attempt to normalise the data based on topography and tree shading, given the dramatic effect these can have on exercise-based sweating, particularly in a hot country. The E4 measures EDA at the wrist which is a compromise to keep the unit relatively compact and discreet. Empatica have subsequently released an updated version of the device which allows measurement at the fingertips, which is generally regarded as a more accurate location, but this was not available at the time of the research. While the standard wrist-based sensor functions quite well, over time the somewhat plasticky wristband can become quite sweaty even in relatively mild climates. In addition, exercise-based sweating has been shown to significantly increase EDA levels in participants, bringing the possibility of a false positive when looking for increased arousal (Posada-Quintero et al. 2018).

During Tess' field study we observed that EDA can be seen to consistently climb for some participants during walks up even quite gentle hills in the mild climate of Birmingham – the additional physical effort also showing up in increased heart rate. We were therefore very cautious to check for the effect of walking up hills when examining data apparently showing increasing participant arousal. Although we were impressed by Shoval et al.'s project, we felt some additional caution needed to be employed in some of the conclusions relating to the E4. As an example, elevated EDA among participants who had just walked up the Mount of Olives on an unshaded path in the Middle Eastern sun might have had more to do with environmental factors than the fact they were visiting a religious site. Again, this reiterates the point that data generated by these devices needs very careful interpretation, particularly when examining participants moving around complex and uncontrolled environments.

Eye tracking

The eye's fovea is at the centre of our field of vision and captures the detail of what we see. Beyond the fovea, which covers just ~2 degrees of our vision, our field of view becomes progressively less detailed, simply picking up on things like shape, colour and movement. The brain processes cues from peripheral vision as a high priority, which means that if we sense something in the periphery that we should be paying attention to, the eye moves immediately in that direction to bring it into the detailed foveal view. When the fovea pauses on a point of interest, we describe this as a fixation, with the eye moving rapidly between fixations in zig-zag patterns known as *saccades*. Fixations can be measured in milliseconds and from these we build up information about the things that appear in front of us. Our eyes *dwell* when we spend time looking at a specific element for an extended period. As the eye is moving during saccades, we are effectively blind, but our brains reconstruct the images gathered during fixations to give us the impression that we 'see' everything within view. If we were to actually attempt to fixate on everything in front of us to construct a detailed picture purely using the fovea, this would take about 16 minutes (Holmqvist et al. 2011). In effect, we only ever look at and see the things we need to at any given moment and our brains fill in the gaps.

This description of the process would no doubt cause a psychologist to cringe and it is the nature of a book of this kind that some of the expert detail is inevitably lost. Nonetheless, it is useful to have at least a basic understanding of what the techniques developed in specific disciplines allow us to do, even if this is simply a question of knowing about possibilities that could be pursued with a skilled collaborator. In this section we will cover two different approaches to eye tracking that we can term clinical and descriptive. Clinical approaches take precisely calibrated quantitative measurements and use these to examine cognitive response. Descriptive approaches are more interested in providing a general account of what participants were looking at in a field of view at any given moment. Both draw on the same technologies but work in somewhat different ways and allow different conclusions to be drawn.

By examining eye movements, we can begin to explore how people consciously and unconsciously build their understanding of the world around them, examining how they look at different elements within their field of vision. We will explore this process more in the case studies discussed below. In terms of the technology used to capture these eye movements, there are two main types of device: screen-based and mobile. Screen-based devices measure where a participant is looking on a screen and allow the researcher to record their gaze – including elements such as fixations and saccades – against the images being shown on the screen at the time. Mobile eye tracking (MET) glasses, conversely, are worn on the head and consist of a front-facing camera which captures the participant's field of view (Figure 3.5). Sensors mounted in the frame of the glasses simultaneously record eye movement so that the researcher can overlay representations of this onto the video of the participant's view. Both systems allow researchers to undertake a quantitative analysis of participants' eye movements in response to different stimuli.

Screen-based devices are more commonly used in clinical-type eye tracking projects, while MET glasses lend themselves more to descriptive approaches although both types of device can be used either way. The technology has advanced rapidly in recent years. Originally, screen-based approaches required participants to be immobilised but today's devices are much more discreet, easy to calibrate and can compensate for a degree of participant movement, so long as they stay in front of the screen. MET glasses, meanwhile, used to take the form of heavy goggles where today they are designed to be as unobtrusive as possible, looking somewhat like a pair of sports sunglasses with clear lenses. MET approaches allow researchers to examine participant's interactions with an actual environment, rather than restricting them to images on a screen.

Eye tracking equipment has always been expensive and remains so even though costs have fallen in recent years. This said, non-clinical-grade screen-based devices are starting to find use in the gaming sector, as a control mechanism for players both on screen and within VR headsets. Where more precision is required, however, there still needs to be a sizeable investment in

Figure 3.5 Tobii Pro 2, mobile eye tracking glasses.
Source: Tobii AB.

equipment and the associated analysis software. As a result, one of the most common applications for eye tracking has been in market research (Pieters and Wedel 2008) and user experience design (Romano Bergstrom and Schall 2014), particularly within the private sector where the analytical insights are worth the significant costs. A fairly straightforward application would be to examine how users engage with a webpage, the amount of time they spend looking at different elements (adverts, text etc.), how long it takes participants to find the part of the page they are interested in and the patterns of visual engagement that lead up to the decision to click on a particular link. This can give crucial insights into the effectiveness of that site. An eye tracking study by Romano Bergstrom et al. (2013), for example, has shown significant age-related differences in how people are able to navigate websites, which has important implications for how these sites are designed if they are intended to serve an older audience.

Research on packaging design using eye tracking can analyse the decision-making that nudges consumers toward picking one product over another. Prompting greater attention in the 'orientation' phase of looking at a set of items can drive an increased likelihood that the product will be more efficiently considered in the 'discovery' phase where buying decisions are made (Husić-Mehmedović et al. 2017). Eye tracking is valuable here because qualitative surveys of product recall miss a lot of the detail around decision-making – bluntly, people do not always know what drove them toward selecting a particular product. Such research does not necessarily have to be about manipulating consumers into buying one product over another, however. Graham et al.'s (2012) meta study, for example, looks at research into the positioning of nutritional labels on food packaging and how these might be better designed to encourage consumers to engage with this information. They highlight how placing nutritional labels centrally on the packaging, reducing the visual clutter around them and listing the nutrients in order of those having the greatest health impacts leads to significant improvements in consumers deriving relevant information from the labels.

Eye tracking does seem to lend itself to collaborative projects as a tool that can shed light on a number of different disciplinary questions. A nice example of this is in the work of the interdisciplinary Eye Tracking and Moving Image Research Group, which is primarily based in Melbourne. Bringing together film scholars and eye tracking specialists, the research questions moved on from earlier, scientist-led studies looking things like quantifying the extent to which viewers focus most of their attention in centre of a movie screen. Instead, as part of an innovative edited collection (Dwyer et al. 2018), they started to examine topics such as the differences in how viewers watched mainstream and experimental movies, how the presence of a star actor concentrates attention to particular parts of the screen, and how viewers become involved in problem solving within murder mysteries. Without a collaborative approach, the nuances of film theory and the subtleties of eye tracking techniques would have been lost and such a collection of essays would have been significantly less innovative and revealing.

A key advantage of screen-based studies is their accuracy and replicability. Because one can show exactly the same imagery to different participants, it is easy to make comparisons between them. Undertaking these studies within a lab setting allows other environmental variables that might influence participants to be controlled. It is also easier to identify specific areas of interest within a consistent set of imagery which can be targeted using analytical software to quantify response across participant groups. As such it is unsurprising that screen-based studies are more common in clinical-type approaches. This is not to say, however, that mobile eye tracking glasses cannot be used in research generating quantified data. A nice example of this comes in a study by Diaz et al. (2017) analysing how trainee cadets and experienced pilots examine their cockpit instrumentation. Participants were sat in a flight simulator meaning that they had a relatively consistent set of objects to look at – the instrumentation and the view of the landing approach out of the simulator windows. Advanced eye tracking software can examine MET video frame-by-frame and match it to reference photographs of an environment such as a simulator cockpit. Thus it is possible, even with people moving around, to automatically quantify how much time is spent looking at different areas of interest in an environment and compare this between participants by connecting it to the same reference photos.

Seeing this image matching in operation is extremely impressive. In the case of Diaz's study, they were able to easily differentiate between participants looking out of the simulator windows and looking at their instrumentation. The headline figure from the study is that experienced pilots only spent 77.5% of their time looking outside the cockpit when undergoing a landing simulation exercise, compared to 85.4% for cadets. Experienced pilots thus could be shown to spend less time checking their position by looking out of the window because they were more confident in gaining the necessary information from their instruments. This in turn meant that they would be faster to pick up possible safety issues shown in the instrument data which would be less obvious to someone concentrating on looking out of the cockpit.

This study demonstrates how eye tracking has particularly interesting applications in training scenarios, showing how experts look at the world in a measurably different way. Learning how to 'see' like an expert can be vitally important, particularly in high-risk environments. In addition to gaze tracking, studies have also examined pupil-response as a measure of cognitive load to examine how trainees cope during different tasks (Rosch and Vogel-Walcutt 2013). Such work has proven to be of tremendous value in fields such as surgery which combines precision and high pressure decision-making (Tien et al. 2014). Cost remains a significant issue, however. A mobile eye tracking package including analysis software can cost more than £20,000, with replacement headsets costing thousands of pounds per time. Thus, there may be training scenarios, particularly in the public sector and high-risk environments, where the costs would be prohibitive. Phil has discussed a potential project with his colleague Dominique Moran who runs the Carceral

Geography lab at the University of Birmingham. As part of her work on building design she has interviewed prison officers who have talked about the way that experienced staff can more quickly assess potential danger in a given situation. Phil and Dominique had talked about the potential of using MET glasses to investigate whether this anecdotal evidence could be quantified and training for new staff designed around these insights. For confidentiality and safety reasons, all the video footage recorded by the MET glasses would have had to remain on the prison site, which would have created practical difficulties with such a project. More than this, however, given the likelihood of the glasses being torn off, stamped on and otherwise damaged in the often chaotic situation within UK jails, Phil and Dominique decided that the project would be too costly and high risk to be worth pursuing.

In this sort of situation, the MET glasses themselves could potentially further inflame tensions within a high-risk environment, provoking disputes with prisoners objecting to being filmed. Even outside the highly regulated yet volatile spaces of a prison, the use of MET glasses does raise considerable questions about the ethics of privacy in public space. Any filming in public needs to be undertaken sensitively and with due attention paid to confidentiality. Ethnographic work using film is sometimes shot from the perspective of the participant, using sports and action cameras which removes participants' bodies from being the *subject* of the research. Arguably, this means that participants instead co-construct the research data *through* their bodies (see, for example, Brown et al. 2008 on off-road cycling). Nonetheless, using such approaches in environments containing other people inevitably records the faces of individuals who have not consented to be part of the research. This puts a significant responsibility on the researchers to ensure that these videos are treated as confidential data unless faces and other identifiable markers are blurred. It also potentially puts participants at risk of being challenged by members of the public objecting to being filmed.

It should also be noted that the eye tracking data adds another layer to issues around participant confidentiality. By revealing elements of decision-making processes, eye tracking data can give insights into an individual participant that they may not have fully understood when giving their consent to be recorded. To take a slightly facile example, one of the other elements of eye movement that can be revealed through this technology is the *smooth pursuit*. This is where an individual tracks the movement of an object through their field of view (Krauzlis 2004). Participants may be uncomfortable with footage that shows, for example, that they tracked the movement of an attractive person walking past them. While this eye movement may not have been underpinned by sexual arousal and could simply be a case of distractedly tracking a moving object, this has the potential to cause embarrassment to a participant if given the opportunity to review their footage.

Market research uses of MET usually take place in relatively confined, indoor spaces, such as asking participants to wear the glasses while walking around a supermarket. Again, the capability of the analysis software to match image targets within moving video is crucial here. Researchers can take a photo of a shelf of products, identify particular areas of interest within that photo and then automatically search the video footage for information about how each participant looked at that shelf as they walked past. The advantage of using MET within this kind of scenario is that participants can undertake the task in a more natural manner, rather than sitting in a chair in a lab looking at pictures on a screen.

Clement (2007) recruited 61 supermarket shoppers, asking them to wear MET glasses while doing their regular grocery shopping. He then analysed product-buying decisions specifically related to purchases of jam and pasta – chosen as staples that feature in many shopping baskets and for which there are many different brands and varieties. This work demonstrated that for customers who were not seeking out a specific product there was no statistically significant difference in how they looked at different items in the same product category until the point where they started to make a decision about the one they intended to buy. This kind of analysis can give really interesting insights about how people behave when shopping, for example, 69% of people looking again at a product on the shelf that they have just put into their basket. Again, these are valuable insights for the design of both packaging and store layout, generated from observations of real environments rather than the artificial confines of a lab.

Similar approaches have been used in museums to examine visitor interaction. Eghbal-Azar and Widlok (2012), for example, asked participants to walk around exhibitions in museums in Stuttgart and Marbach to examine whether the researchers could identify a common way of interacting with these exhibitions, which they termed a 'museum script'. The study is interesting for the way in which it combined a descriptive approach to MET with post-visit interviews, asking participants to reflect on their experiences in the exhibitions while watching back the video of their visit overlaid with a representation of their eye tracking data. They concluded that, taken by itself, eye tracking data can be ambiguous, since you do not necessarily know why participants were paying attention to a specific object nor why they passed through the exhibition via a particular route. Here we see similar arguments about the importance of a mixed methods quantitative and qualitative approach that we noted in the section on physiological measures above.

We can also see the potential to design research interventions around user interactions with different spaces. A study by Tatler et al. (2016) examining visitors to the historic visitor attraction of Owlpen Manor in Gloucestershire examined whether expertise could alter interactions with the space. The manor is famed for its painted cloths and the research team took a group of tapestry experts to the site and examined what they paid attention to when passing through the manor. They also experimented with groups of non-experts, some of whom were briefed about the historical significance of the painted cloths and some not. Perhaps unsurprisingly, the non-experts spent much less time looking at the painted cloths than the experts, although those who had been briefed about their importance did pay them greater attention during their visit to the manor. This kind of study gives important insights into how educational activities can be enhanced to improve the visitor experience.

Participants wearing MET glasses in public can become very self-conscious – a limitation that Eghbal-Azar and Widlok themselves note – and this was a particular problem with older studies where the equipment used was considerably larger than today. Not many participants would feel particularly comfortable walking around in public wearing oversized goggles with a camera mounted to them. Although modern devices are much more discreet this brings its own problems because passers-by can react badly to someone 'secretly' filming them. In relatively small, indoor environments like a supermarket or a museum it is much easier to generate informed consent among non-participants, or at the very least warn people that filming is taking place, simply by putting up signs at the entrance. Doing research in larger, outdoor environments thus becomes more challenging.

As always with research there are trade-offs to be made here. Using MET glasses in an outdoor environment can create safety and privacy concerns. It is also much harder to automatically undertake a quantified analysis and outdoor environments bring a range of potentially confounding factors (time of day, weather, participant fitness etc.) that makes comparison between participants less reliable. In assessing these trade-offs Jodi Sita (specialist in neuroscience and anatomy) and Marco Amati (an urban planner) worked on an interesting project using a screen-based approach to examining and outdoor environment (Amati et al. 2018). All 35 participants in their study watched videos of 'walks' through different urban parks in Melbourne. Using an automated classification algorithm, each frame in the video footage was categorised into different areas of interest, paths, benches, bushes, other natural features and so on. These areas of interest were then analysed against participants' eye tracking data to measure what they were looking at on screen during different stages of the experiment. The virtual walks were of two parks, one lush green with tall trees and dense vegetation planted in a European style, the other somewhat parched and more sparsely planted with native Australian species with pale green and yellow colours dominating the view. Participants spent much more time looking at human-made features (the path, benches etc.) in the park with more Australian species than in the park with European planting where they paid more attention to the surrounding vegetation.

Similar to the Owlpen Manor example, it is possible that teaching a participant more about the species that have been planted in a park might make them more inclined to spend time looking at them rather than concentrating on where they were walking. This would be an interesting hypothesis to test with a repeat study. Regardless, this kind of project gives really valuable insights for planners and those working on the design of urban parks. Of course, there are limitations here. Sitting in front of a screen watching a video is a very different experience to actually walking in a park. Had they used MET glasses for the project, however, there would have been a huge number of potentially confounding factors to take account of like time of day, weather, the age and health of the participant and so on. It would have also been harder to use the visual algorithm to automatically identify the different elements of the landscape participants were looking at.

The trade-offs between lab and field-based studies can be seen in the work of Pieter Vansteenkiste's lab in projects examining urban cycling. In a study looking at the effects of cycle lane surface quality on cyclists' attentiveness they undertook a mix of field and lab research. Participants rode a 4km route on four different road types around Ghent in Belgium while wearing a head-mounted eye tracking device (Vansteenkiste et al. 2014). The basic conclusions were that a cycle path of a poorer quality *does* distract the attention of the riders more, so that they need to pay more attention to the road surface than to scanning for other potential dangers. In a follow-up study they found this to be fairly consistent across both children and adults, although children generally were found to have lower skills in anticipating potential hazards (Vansteenkiste et al. 2017). They also experimented with seeing whether they could replicate the field measurements in a lab setting (Zeuwts et al. 2016). Participants from the Ghent study were invited to a lab 11 months later, riding an exercise bike in front of a projection of the video they had originally recorded. The study demonstrated that some of the differences of attention between high and low quality cycle routes could be replicated in the lab but the team noted that 'centrality bias' (where people tend to concentrate their gaze in the centre of a screen) had an impact on the effectiveness of the lab-based work.

Clinical approaches to eye tracking can be seen very much as the gold standard approach and require significant expertise to generate meaningful insights. In recent years, user-friendly software has become available that produces compelling visualisations (such as heatmaps showing where gaze is concentrated) requiring less skill to produce and interpret. As a result, there is some interest in using this technology in a way that experts would dismiss as broadly descriptive – simply identifying what participants are looking at and then interpreting what this might mean. Although such work lacks the precise information about cognitive behaviour that can be derived from clinical approaches, it can nonetheless highlight interesting issues that might be explored in more detail through follow-up studies, potentially working with more skilled collaborators.

Eye tracking is also an emergent technology within VR applications. Eye tracking enables a process known as foveated rendering (Patney et al. 2016). VR headsets contain two small screens, one for each eye. The higher the resolution of these screens, the better the sense of immersion but the more computing power that is required. Foveated rendering allows the computer to only produce high levels of detail at the points on the screen where the user is fixating, thus saving a great deal of processing power while allowing higher quality images to be perceived by the user. The same technology can also be used by researchers to combine the best qualities of both MET and screen-based approaches. Automatically identifying a book or a jar of jam in a video requires quite complex computer analysis that will not be 100% accurate. In a virtual environment, every object has an independent existence which is known to the computer. As a result, eye tracking software used in VR allows researchers to quantify in real time how every object is visually interacted with by a participant wearing a headset. One can very precisely calculate things like dwell time or number of fixations on different objects within that virtual environment across a large participant cohort with minimal processing time (Tobii 2019).

One of the most significant non-research applications for eye tracking has been as a control mechanism within gaming (we will discuss the use of gaming technology for research in more detail in the next chapter). Non-clinical-grade devices have been made available that simply clip to the bottom of a PC screen – indeed, some gaming laptops now come with eye trackers built in. When playing a game such devices are commonly used to control things like the game's camera angle – a player glancing to something on the side of the screen will cause the camera to move so that the object is transferred to the centre of the field of view. They can also be used to prompt game activity, selecting items by looking at them or having a character react to being stared at by the player. One of the first mainstream games to support the technology was 2014's *Assassin's Creed Rogue* (Webster 2015) and at the time of writing there are 116 games supporting this technology available via Steam, the main online store for downloading PC games.

The Swedish company Tobii is one of the leading developers of eye tracking technology and it has provided some tools for accessing the outputs of its games-focused eye tracking devices. This gives an opportunity for non-experts to explore some of the potential offered by this technology. Tobii provide a number of free tools for software developers, from high-level integration with the Unity games engine to simpler plugins to create eye tracking overlays. Phil has given some of his students the chance to develop small research projects using a Tobii 4C gaming eye tracker. They employed Tobii's screen capture software, which was intended for use in games streaming but can be repurposed for use with videos, photographs, web pages and so on. Participants do not see the representation of their eye movements when looking at the screen but it appears as either a dot or a heatmap within the recorded screen capture. Any analysis is, of course, purely descriptive and Tobii explicitly prevent users from hacking the device to output numerical data. One can cynically speculate that this restriction is because for many researchers the device would otherwise be 'good enough' to generate useful quantified outputs, thus undercutting the market for Tobii's clinical-grade devices and software.

Phil emphasises to his students that this approach only allows an impressionistic, descriptive analysis to be undertaken. Some have chosen to try combining this with the use of an Empatica E4 to explore whether any connection could be observed between physiological stimulation and eye tracking. This combined approach is common in eye tracking studies (for example Hurley et al. 2015) although here was only undertaken in a very basic manner. It did, however, lead to an interesting group discussion about potential ethical problems. One group were keen to examine how men and women viewed sexualised advertising materials differently. One risk with such a project would potentially be inadvertently outing a closeted gay participant who was shown to be respond more to same-sex imagery. Again, this reiterates the difficulties around consent in these kinds of projects where participants may not really understand what inferences can be drawn from capturing their bodily reaction.

Conclusion

We end this chapter by reflecting on a point made at the close of the previous one – most people have no idea just how much researchers can find out about them by monitoring their bodies. Critical scholars therefore have a significant responsibility to ensure that *informed* consent is not just an empty phrase written into ethical review documents, but an ongoing process to give participants a greater degree of ownership over the data collection process. It is easy to get lost in the enthusiasm for what new technologies allow us to do and lose sight of our participants as *people* rather than mere numbers.

Nonetheless, the range of technologies which have become available to measure and alter the body do provide some incredibly exciting possibilities for researchers. We would summarise these as being covered by four broad themes:

1   projects that work with participants to examine the data being produced by the smart devices they routinely use and the ways in which they modify their behaviours and bodies;
2   projects that use different monitoring devices to explore participants' interactions with different scenarios both in the field and the lab;
3   projects that collaborate with experts in the use of a particular technology in order to develop novel, interdisciplinary insights;
4   projects that employ these devices within user testing with a view to improving the experience of engaging with different products and spaces.

Simply trying out some of the different devices that have become available can be a fascinating experience as it highlights new directions in which research can be taken. A gaming eye tracker simply will not produce data that would meet the standards necessary for a more clinical study, but it can point the way toward what can be done with more advanced technology. Trying this out with a purely descriptive study as our students have done, opens the door to a conversation with experts in this area.

Critical researchers are not going to invest tens of thousands in a piece of clinical-grade equipment that they do not know how to use just to play around with it. There is, however, a temptation to buy some of the cheaper pieces of equipment – such as the E4 wristband that we have used – which have the potential to generate very high-quality data. This of course brings us back to the key issue of how to analyse and interpret these datasets. When working with participants there is a danger that they might expect researchers to be able to flag up any health problems shown by their data, meaning that there are issues around managing expectations. Just as significantly, devices like the E4 or the EMOTIV EPOC+ work best in a lab setting where environmental factors can be controlled, making it easier to isolate the impact of the particular stimulus being tested. Once out in the real world, there are a host of potentially confounding factors to take account of.

Because of this we have to be really careful about the conclusions we draw; mixed methods approaches are therefore crucial to help us explain some of the patterns we see in the physiological data. We also need to be cautious when making claims about being able to read emotional response through these devices – the circumplex model is, after all, just a *model*. What we get are inferences about emotional state which need to be contextualised. This kind of caution takes us further away from some of the more heroic claims that are made about bodily quantification. Being able to quantify how much and what kinds of exercise we take can be useful but wearable devices should be seen as complicit within a broader neoliberal discourse of making the individual responsible for their health. Tacking the obesity epidemic in developed countries requires a more fundamental rethinking of how our societies operate but it is easier to blame the individual for being overweight and have doctors prescribe the use of a step counter.

Of course, wearable and implanted medical devices have transformed the lives of some patients and this is a cause of justifiable celebration. While fitness devices can be incredibly valuable for professional athletes, in the consumer market there is a tendency toward pandering to the worried well and marketing magic bullets to those with stressful, difficult lives. The biohacking movement takes this approach to its extreme suggesting that the body can be optimised and augmented through technology-led interventions. Some of these claims are, at best, dubious. There is a clear gendered component to this, with a masculinist disavowal of the fragility of the body. In the biohacking discourse the body becomes an object to be manipulated but treating the body as property highlights questions of who owns and controls it.

Biohacking is entangled in an approach to disrupting traditional types of scientific research and authority. This can be framed as being empowering for vulnerable groups, such as reducing the dependence of the trans community on the medical establishment. But it also brings us back to questions about ethics. The US Government Accountability Office concluded in 2010 that home genetic testing was 'misleading and of little or no practical use' (quoted in Udesky 2010). Knowing that one is of slightly elevated risk for Alzheimer's or macular degeneration in later life is not necessarily useful information for the individual but it can create undue worry. Again, this brings us back to having the capacity to meaningfully *analyse* the data that new techniques allow us to gather about the bodies of ourselves and those of our research participants.

Critical social science and humanities scholars are always going to be cautious about the kinds of disruptive approaches championed by enthusiasts for movements such as garage biology. Certainly, as researchers, formal processes of ethical review prevent us from disrupting things in such a way that we risk putting participants at harm. Working with devices that measure aspects of participant physiology brings these questions into sharp relief. Nonetheless, despite all the caveats, there are some genuinely interesting opportunities to pursue when engaging with the technologies being developed in this area even if this merely takes the form of exploratory or pilot work for a future interdisciplinary collaboration.

4   Gaming and virtual landscape

Introduction

In 1983 Bug Byte Software released *Manic Miner*, a fiendishly difficult platformer for the ZX Spectrum. Received to great critical acclaim in the emerging field of games journalism, it quickly became one of the best-selling titles for the new computer. Players had to dodge a variety of lethal contraptions while collecting items strewn around a series of screen-sized 'caverns'. The joy of the game for players in the early 1980s was *exploring* this exciting landscape, where the slightest mis-step could send the game's hero to an early death.

Manic Miner and its sequel *Jet Set Willy* (1984, Software Projects) were written, designed and coded by Matthew Smith, who was just 17 when the first game was released. Smith was, in many ways, the archetype of the bedroom coder, a group of men who play a large role in the origin myth of the UK's gaming sector. While examples can still be found of hit games produced by a single person – Dong Nguyen's 2013 *Flappy Bird* for example – this is now quite unusual. Today the sector is highly complex and collaborative, with writers, actors, designers, musicians and coders coming together to produce titles across a range of genres and platforms. Larger studios have workforces numbering in the thousands with offices spread across the globe. Gaming is one of the most lucrative elements of the creative industries, with the sector worth an estimated $137.9 billion in 2018, 47% of which was the rapidly growing area of mobile gaming (Newzoo 2018). Clearly, this is a sector with tremendous cultural and economic reach.

It is possible to equate an early game like *Manic Miner* to pioneering work in cinema, such as George Meilles' 1902 *Le voyage dans la lune*; Matthew Smith and his peers were creating the basic language of a new medium. Tropes of hazard, conquest, death and the male hero run deep in gaming which creates points of tension for critical scholars engaging with this area. Meanwhile, the sheer level of detail in the landscapes of contemporary games, which often comes as a shock to non-players, offer rich opportunities to explore questions of embodiment within a range of virtual spaces. In many ways the boundaries between virtual and material worlds are consciously blurred by gaming. Emotional, physiological and sensory stimuli are used to give players a sense of *presence* in virtual landscapes. There are also significant questions for researchers engaging with these spaces in terms of how socially problematic behaviours can be reproduced in virtual environments and the kinds of physical and embodied skills required to access these worlds.

This chapter explores the ways in which scholars have used games, not merely as texts for analysis, but as case study sites within which to undertake research via digital ethnography and active interventions with participants. The chapter breaks into four parts, firstly examining the origins of games studies and some of the specific cultures that have arisen around gaming. Secondly the chapter turns to explicitly examine how embodiment is dealt with in games. Thirdly, the creation of digital landscapes is explored, specifically in how we interact with these digital spaces. Finally, the chapter reflects on the research possibilities that are opened by scholars building their own game environments.

Games studies and gaming cultures

Aarseth (2001) claimed that the launch of the *Games Studies* journal in 2001 represented the emergence of a new interdisciplinary research area. Arguably, however, games studies as a research field has been a victim of its own success – as it helped to legitimise games as a topic of research, so scholars have increasingly felt comfortable retreating to their own disciplines to discuss them (Deterding 2017). Despite its interdisciplinarity, games studies is closely aligned to a cultural studies tradition, with a predisposition toward using discourse analysis to examine games as texts. As the field was being established this led to a somewhat inward-looking debate about the relative merits of narratology versus ludology – analysing story versus analysing play (Frasca 2003). This debate remains a useful reminder, however, that games are very different to traditional media texts such as novels and cinema because they have a high level of interactivity. Story and play style are deeply intertwined in games and so it makes sense to study these together, acknowledging that the game text being examined is partially co-constructed by the player.

The subtleties of questions around ludology continue to animate discussion in some quarters of games studies (for example, Vargas-Iglesias and Navarrete-Cardero 2019 in press). In many ways this should be unsurprising because *play* remains vital to our understanding of games; no matter how interesting the story, if the gameplay is clumsy or buggy or frustrating, players will not engage with a given title. As a result, Chapman (2016) notes a serious tension between narrative and ludic worlds. Where games are trying to deal with serious themes such as 'realistic' recreations of actual wars, the logics necessary for enjoyable *gaming* do not sit well with what actually happens in combat situations. Games are often simply win or lose narratives, leaving little room for ambiguity, brutality and the stories of those who are not the conquering heroes:

In a WWI game of this type, wherein the player controls only a single historical character, the death of this character means that the player must either be able to reload – potentially disturbing the serious historical narrative … – or be denied the opportunity to reload and replay, disturbing the typical videogame experience.

(Chapman 2016, np.)

As a result, interesting story or social commentary elements are almost scribbled in the margins of many games so as not to distract from a playable text which is designed to be *fun* for players. Longan (2008), for example, reads adventure game *Escape from Monkey Island* (2000, LucasArts) as being a critique of gentrification in contemporary cities; pirates are banished from Melee Island by the game's villains in order to turn it into a more capitalistically productive resort destination. Even a title like the rebooted *Doom* (2016, id Software) has an anti-capitalist satire at its heart. *Doom* has a refreshingly uncomplicated play style which might be summed up as 'see demons, shoot demons'. In its cut scenes (cinematic breaks from the play action) the excuse for why the protagonist is running around killing endless demons is revealed to centre around a greedy corporation seeking to tap into demonic energy in order to generate cheap electricity. This scheme has gone horribly wrong, freeing demons into the human world meaning that the hero is tasked with resolving capitalist folly by eradicating the demonic scourge. In truth, however, this anti-capitalist satire is not what most people are thinking about when they play the game; they simply progress from room to room, merrily slaughtering demons as they go.

Beyond this tension between story and play, perhaps the most significant concern facing researchers interested in gaming are questions of culture and identity. Gaming as a medium undoubtedly has a gender problem – as well as issues around sexuality, ethnicity and other intersectional identities. The cliché of the gaming audience being dominated by teenage boys is simply not true, however. Industry lobby group the Entertainment Software Association (2018) points out that the average age of US gamers is 34, with the number of female players over 18 being nearly double that of male players under 18.

Nonetheless, gaming culture remains uncomfortably sexist, something which starts at the level of production. Despite some 45% of US gamers being women, the industry itself is still dominated by men which has a significant impact on the kinds of stories deemed worthy of being invested in, the kinds of playstyles employed and the cultures tolerated/encouraged among players themselves. Analysing tweets sent as part of the #1ReasonWhy conversation, Ochsner (2019) identified that women who had left the gaming industry cited the fact they were judged on different standards to men, did not have the same recognition and status as male colleagues and all too often simply did not have their voices heard.

The disadvantaging of women in the gaming industry mirrors that in the wider tech sector. The lack of female voices within the industry fosters a culture which, among other things, has a tendency toward treating women as disposable props within game landscapes. Back in 2006 it was still possible to undertake a content analysis of games journalism on the Gamespot website and find that only 33% of reviews mentioned an active female character as against 75% mentioning an active male character. Likewise, of reviews that pictured a female character, 41% contained sexually suggestive imagery as against only 4% where the review pictured a male character (Ivory 2006). While the representation of women within games became more nuanced in the decade that followed, this ran alongside the rise of a highly vocal, toxic backlash by some players opposed to this toning down of sexist game design.

The backlash reached fever pitch with the Gamergate controversy of 2014. The hashtag #gamergate was coined by actor Adam Baldwin in a tweet that he subsequently deleted. Its defenders argued that the movement started because of concerns around ethics in gaming journalism, with allegations being made that the relationship between developers and journalists had become too close. The catalysing event was indie developer Zoë Quinn being accused by an ex-boyfriend of dating a journalist in order to receive a positive review of her newly released game *Depression Quest* (2013, The Quinnspiracy). Although these claims were quickly and easily debunked, the Gamergate movement rapidly morphed into a wider attack on any perceived attempt to diversify gaming away from catering solely to a male, white, cis, straight audience. Quinn and commentator Anita Sarkeesian became particular hate figures within the Gamergate movement, receiving multiple rape and death threats. Sarkeesian was targeted because she had received crowdfunding to produce her 'Tropes vs. Women in Video Games' YouTube series (2013) examining sexist assumptions within games. Drawing on a deep knowledge of games and gaming cultures, to feminist and allied scholars the videos would doubtless seem entirely uncontroversial, making some convincing points about problematic attitudes to women in games. Gamergate propelled what would otherwise have been a somewhat niche set of videos into much greater public consciousness, surfing a wave of bile among those who perceived Sarkeesian's call for greater diversity as a threat.

Games scholar Torill Mortensen (2018) compares Gamergate's activists to football hooligans, using their leisure activity as an excuse to get involved in a fight, rather than representing a coherent movement with a meaningful ideology or agenda. The key tactics used were generating a constant stream of abuse via Twitter and other online platforms and, more chillingly, the practice of 'doxing' – the publication of personal information. This information could include home address and the names of schools attended by the target's children. The journalist Leigh Alexander was a focus of doxing and other attacks after she wrote an article disowning the label 'gamer' because of its association with the toxic masculinities exposed by Gamergate (Alexander 2014). Games researchers were themselves abused, with Digital Games Research Association president Mia Consalvo and other female scholars identified for criticism – Mortensen was herself doxed on the 8chan messageboard.

This last point has real methodological significance. If games scholars are concerned that they will be abused and threatened for research that challenges toxic gaming cultures, there is a real risk that they will be reluctant to work in this area or will avoid specific topics within it. The clearest example of this is in online multiplayer gaming. The anonymity afforded by online gaming, combined with a timidity by developers to crack down on antisocial behaviour by players, frequently leads to some truly vile in-game commentary directed at other players, going far beyond good natured 'trash talking'. Some games are particularly notorious for their toxic cultures, for example *League of Legends* (2009 onward, Riot Games). Reflecting on her identity as a lesbian gaming scholar, Humphreys (2019, 839) comments:

> As a researcher in an MMOG [massively multiplayer online game], dealing with constant homophobic commentary from various player cultures that I have participated in has also been challenging and ultimately caused me to switch out of some games. You start to ask yourself why you aren't researching something else, less personally offensive and confronting.

Humphreys' experience is by no means unique. Cote (2017) notes five coping strategies employed by female gamers in multiplayer spaces: camouflaging their gender; avoiding playing with strangers; highlighting their skill and experience; being aggressive in response to attacks; and simply leaving certain online games. Researchers working in these areas may feel the need to deploy similar coping strategies to protect themselves from abuse. Our ability to approach research in this area is thus shaped and potentially curtailed by risk to the researcher.

The fact that a major sector of the entertainment industry is complicit in a fan-culture that actively drives women away from engaging with certain products is both culturally shocking and economically bizarre. Of course, scholars can and have researched these toxic cultures within gaming. Christopher Paul (2018), for example, has examined how problematic notions of 'meritocracy' pervade gaming cultures. He has undertaken an analysis of how the mechanics of gameplay are carried over between titles, giving a sense of superiority among those more familiar with these gameplay logics – that they are somehow inherently more *worthy* to play these games.

There is a danger here, however. Games scholars are themselves often committed players which brings a tendency to be drawn toward the richer and more complex worlds of console and PC games, rather than the smaller and less graphically sophisticated games available on mobile phones that are often dismissed by 'hardcore' gamers as being for 'casual' players. Women and particularly older women dominate the so-called casual market, however, and by paying less attention to this area, scholarship risks reproducing some of the more problematic gendered assumptions within the wider gaming industry and community. This is also a research significant gap in terms of understanding the industry not least because mobile/casual games are growing at a *considerably* faster rate than PC and console games.

Writing this chapter as a white, straight, cis man makes me acutely conscious of Humphrey's (2019, 834) critique that games scholars have tended to spend too much time researching the so-called 'triple-A' games released by large studios. These are generally the most risk averse in terms of story and gameplay and most targeted at the hardcore, traditionally male, gaming audience. Large parts of this chapter draw on examples from these blockbuster games and this undoubtedly reflects my own position and interests. This does, however, remind us of the challenges being posed when making the decision about which games and communities to focus attention on within our research and choosing appropriate methodologies to respond to these.

Gaming and the body

Gaming sits at a blurred boundary between the virtual world and our material embodiment. At its most extreme, this blurred boundary is used to claim that gaming *harms* its players and wider society, finding expression in the periodic moral panics that are directed toward games and gamers. Since the 1980s some have framed videogames as having damaging cognitive effects upon players, with a frequent link made between real world violence and the desensitising effects of on-screen shooting and gore. There is, of course, a long tradition in conservative reaction against new forms of activity that are popular with young people. In the 1920s and 1930s, for example, there was genuine concern in some quarters that depictions of sex and violence in the cinema were leading youth astray and undermining the very foundations of society (Rapp 2014, Springhall 1998).

So persistent have been claims about the violent effects of games that Elson and Ferguson (2014) published a meta-analysis reflecting on 25 years' worth of studies attempting to demonstrate a link. They concluded that the evidence linking games with violence was *at best* mixed, with studies hampered by small sample sizes and editorialising that was not supported by the data. Unsurprisingly, much of the work looking at the cognitive effects of gaming has emerged from psychology, generally employing detailed statistical analyses of large-scale questionnaires, or small experimental studies using brain imaging and other advanced physiological techniques. The findings from this work can be very interesting and sometimes counter-intuitive. Zendle et al. (2015), for example, built two versions of a game where players had to run over pedestrians with a car – one with more realistic graphics and one with a more abstract design, but with exactly the same gameplay. As it turned out, participants who played the version with higher quality graphics were *less* likely to access aggressive emotions when tested immediately after exposure than the group who played the less detailed version. Studies like this give an empirical basis from which to challenge popular notions about the effects that games have upon the minds of players.

Some studies have gone further when attempting to counter the narrative linking gaming with violence by looking for positive cognitive affordances generated by gaming. Indeed, the games industry itself attempted to capitalise on this with a series of games designed to improve cognitive function, such as *Dr Kawashima's Brain Training* for the Nintendo DS (2005, Nintendo SPD, released under various regional titles). Boot et al. (2008), tested different cognitive tasks between a control group, experienced gamers and a non-gaming group given 10 hours of gameplay activity. Their study found no real improvement in cognitive function through the training exercise and they suggested that there appeared to be little cognitive value in buying brain training games. Experienced gamers were found to perform better on some of the perceptual and attentional tasks but the authors suggested this was more likely an effect of self-selection – those with greater inherent capability in these areas would tend to become more enthusiastic gamers. Even where cognitive enhancement can be attributed to gameplay activities, this does not necessarily transfer to the material world. Richardson et al. (2011) for example found that regular players were able to more accurately navigate *virtual* 3D environments than non-players but that this did not translate into an enhanced capacity for real world navigation.

Richardson et al. suggested that one possible explanation for this lack of transfer effects between virtual and real navigation skills might be that gaming merely enhances visuospatial mechanisms, where real world navigation requires multisensory engagement with an environment, including the kinesthetic and vestibular. Gaming is not simply a visual experience, however, with considerable design work going into the multisensory qualities of games, including physiological response. Gaming is not purely sedentary, with players having physical reactions to games, particularly elevated heart rate and electrodermal activity during high stress events. This was as much true of early games like *Pacman* (Namco, 1980) where you were running away from ghosts, as it is of the modern immersive 3D first-person shooter. This provides some interesting opportunities for research experiments examining the body-as-interface to the virtual world. Gilleade et al. (2005), for example, created a simple missile defence game controlled by the player's heartrate. The lower the heart rate, the more accurately the missiles could be fired. The stress of failure tended to make participants' heartrates rise further and thus staying calm was key to successful gameplay.

Perhaps the most striking connection between embodiment and gaming comes in the degree of skill required to learn how to play the game in the first place. David Sudnow's (1983) early autoethnographic study reflects on the time and commitment required to become expert at even the simplest of games – in his case the classic bat-and-ball game *Breakout* (1976, Atari). The need to develop expertise is even more acute in the modern multiplayer games that have become the basis of the esports community. Much of the purpose of gaming is in testing an embodied skill: to time a jump, navigate an environment, solve a puzzle or kill an enemy through the manipulation of a controller connecting the player to the virtual world. These actions can best be understood through the researcher actually playing the game – 'playing research' as Aarseth (2003) puts it. Reeves et al. (2009) discuss the need for researchers to develop at least a basic level of expertise in the game that they are researching – either to examine the game itself or the communities that have formed around it. By having an elementary grasp of what is physically required to play a game in terms of hand-eye coordination, researchers can more meaningfully engage with highly skilled players. Reeves et al. have used this approach to analyse embodied skill through a series of vignettes, carefully pulling apart the visual and audio cues, as well as past experience, shaping players' gameplay in the first-person shooter *Counter-Strike* (1999 onward, various developers).

The esports sector has grown rapidly as gaming streaming platforms such as Twitch, YouTube Gaming and Mixer have created global audiences in the millions for events featuring the best teams and players. Witkowski (2018b, Witkowski and Manning 2019) undertook embedded ethnographic research in 2010–12, just at the point where streaming was driving an increasing professionalisation of esports teams. Her work highlights the great difficultly, methodologically, in examining the embodied expertise of the players she worked with noting that '[b]ody memory is sensuous, and often not revealed in talk or just from spectators of play' (Witkowski 2018b, 37). Working with players operating at the highest levels of skills, Witkowski argues, requires an investment of time attending training sessions and undertaking long-term observations of playing practice, alongside researcher play and conventional interviewing. In this sense researching professional esports teams has significant similarities with analysing the embodied actions of elite players in conventional sports.

Researching esports is not only highly time intensive, therefore, but it also requires a fairly advanced level of embodied gaming skill by the researcher to gain meaningful insights into practice. Fortunately, not all research into the multisensory within gaming requires this level of advanced physical capability on the part of the researcher. Music and sound design, for example, can be experienced with novice-level skills yet are crucial to creating the atmosphere of a game, having a profound impact on players' sense of immersion. As an example, running through the streets of 18th century Paris in *Assassin's Creed: Unity* (Ubisoft, 2014) the player hears snippets of French, sometimes angry, sometimes friendly, occasionally in the roar of the revolutionary crowd. This adds considerably to the sense of *being in* France during that period, but the illusion is shattered when encountering the game's many thugs and soldiers who, for reasons that the game designers never convincingly explained, shout in cockney-accented English (Lewis 2014). This indicates which characters can be interacted with, but players must learn to ignore the jarring effect of these uncanny audio cues which reduce the sense of being immersed in an historic landscape.

Just as in film and TV, sound design and particularly music act as an emotional intensifier within games. Unlike film, the length of time a player engages with a give task is not fixed which creates a challenge for producing music that, for example, is fast paced during a fight, but then becomes much calmer once the danger has passed. Musicians thus create soundtracks that can be broken into parts, to create cues of different lengths. One of the most interesting examples of this came with the procedurally generated space exploration saga *No Man's Sky* (2016, Hello Games). Sheffield-based band *65 Days of Static* worked with the developers to allow fragments of music to be assembled algorithmically by the game itself in response to what was happening at any point in the adventure (Seppala 2016). Genre of music can be crucial here. The *Mirror's Edge* games (2008, 2016, EA DICE) have parkour as their primary mechanic, and the sound design is a mix of minimal techno and trance where the beats are enhanced as the protagonist runs across rooftops, mimicking the increased heart rate and excitement of undertaking such activity for real. Techno is a particularly useful genre for this as it is more easily broken into parts so that the 'song' can come to an end more naturally when the player reaches the end of an action sequence.

The purpose of aligning visuals, sound and physiological response as a multisensory experience is to dramatically enhance player satisfaction and sense of immersion. Csikszentmihalyi's (1990) notion of 'flow' has been extensively used within videogame scholarship to explore how game designers can optimise players' satisfaction. Csikszentmihalyi argued that people find happiness where they reach an optimum state of psychological immersion in a task, becoming so absorbed that they are not thinking about anything beyond the task. Too easy and a task becomes boring, too difficult and it can generate anxiety. As an individual's skill level increases, so the task needs to become more challenging in order to maintain the balance between boredom and anxiety.

Balancing boredom and anxiety to generate flow is a key task for game designers and leaves open some interesting opportunities for research employing different kinds of play. de Peuter (2015) identifies three forms of counterplay within gaming: playing games in ways for which they were not intended; holding protests that disrupt existing games and; creating new kinds of games that open up discussions beyond traditional game themes, especially around social justice. Counterplay experiments have proved quite popular among my final year undergraduates as a means of revealing the ways in which game designers attempt to channel behaviour toward particular activities, even in nominally 'open world' games. I asked one of my undergraduate groups (Eoin Byrne, Thomas Strakosch and Cealladh Sullivan-Drage) to recruit participants to play open world driving game *Forza Horizon 3* (2016, Playground Games). The *Horizon* series allows players to drive a variety of cars across different landscapes, urban and rural, both on- and off-road. Participants were asked to spend five minutes driving as if they were in the real world – keeping to speed limits and obeying traffic rules – and then to have five minutes driving however they wanted. In order to enhance the sense that this was a substitute for a real driving experience, participants were given an Xbox steering wheel and foot pedals to use, as well as a drivers' eye view from the car's cockpit.

Participants found the counterplay task quite frustrating because it broke the game's illusion of being a realistic representation of the driving experience by highlighting, for example, that you cannot turn your head to look for oncoming vehicles at junctions. For all that the game landscape seems to be a reasonably accurate simulacra of actual city streets, attempting to treat this as a real landscape quickly reveals that it is not. Everything about the *Forza Horizon* series – the lack of stop-start traffic, the wide, open roads, the absence of police and lack of consequences when crashing – draws the player into driving as quickly as possible. As a result, driving normally in a game landscape feels quite strange. To many participants it almost came as a relief to reach the second phase of the experiment where they could drive as recklessly as the game encouraged them to even though, if this were the material world, such activity would have been considerably more stressful.

Virtual landscape

As computers have become more powerful, storage cheaper and more plentiful, so the complexity of games has increased dramatically, both in terms of graphical sophistication and the sheer scale of the virtual environments available for players to explore. Geographers conceptualise landscapes as being a co-construction between an environment and the perception of someone within that environment. We can, therefore, speak of virtual landscapes existing within games as a combination of the virtual environment built by the developer and the ways in which an individual player interacts with that environment. It is now more than 30 years since Cosgrove and Daniels (1988) extended the analysis of landscape into an analysis of landscape *representations* (such as paintings and maps) meaning that it is not such a great conceptual leap to examine games through this kind of lens. The conceptual and methodological tools developed by geographers to examine landscape can thus be usefully adapted to analyse the ways that players engage with game environments.

Landscapes are not value neutral but reflect power in society. Rose (1993) subsequently extended Cosgrove and Daniels' Marxist analysis to reflect on how gendered assumptions underpinned our understanding of landscape. Given the discussion of gaming cultures above, this gendered analysis remains highly relevant when examining virtual landscapes. One of the critiques that Rose makes is that conventional landscape analysis ignores the role of male pleasure in the construction of the gaze, and she reminds us that the act of looking is an act of consumption. This gendered gaze places the female body as yet another object to be consumed in the construction of a passive landscape rendered pleasurable for a male viewer.

The position of the game camera is significant here. In first-person games, the view on the screen is designed to put the player behind the eyes of the character they are playing. Third-person games, conversely, place the player's character in front of the camera as yet another object within the game landscape. It is no coincidence that playable female characters are more common in third-person games, creating an object of aesthetic pleasure for a male audience to consume while playing. The character of Lara Croft in the *Tomb Raider* series (1996 onward, various studios) is perhaps the most iconic example of lead character as eye candy within the game landscape.

Critical scholars are acutely sensitive to tropes of conquest, yet exploration, subjugation and mastery are at the heart of many games, though without meaningfully depicting the material implications of such activity. Chapman, for example, has written extensively about the representation of historical events and spaces within games. He notes that because gaming is a relatively immature medium, it is still possible to take a near complete sample of games depicting a particular time in history; in one study, for example, he examined 58 published games that depicted World War I (Chapman 2016). His analysis revealed that the sample was dominated by games depicting aerial combat arguing that both the emotionally sensitive subject matter and the actual *mechanics* of play lend themselves to a distancing from the horrors of trench warfare. A large-scale survey of this kind necessarily involves reducing each of the games to a set of dominant tropes. Nonetheless, by looking at a range of themes and playstyles across an entire genre in this manner, some of these broader issues are thrown into stark relief.

Given that conquering landscapes is such a dominant theme within gaming, games studies has made surprisingly little use of postcolonial approaches. An important exception is the work of Mukherjee (2017, 2018):

> Imagine a Iraqi gamer playing *America's Army* (2002) or a player from Zaire playing *Far Cry 2* (2008): The game's rules constrain him or her to follow certain assumptions about his or her culture that he or she, being marginal to the identity the game constructs, is unable to protest.
>
> (Mukherjee 2018, 511)

As the games sector has evolved rapidly, there is an opportunity for scholars to take a longer-term view of how different cultures are represented in the history of gaming. Chapman's approach of looking at large samples of games within a specific genre thus has real merit here. There are interesting projects to be done using postcolonial theory to examine how, for example, representations of Africa have changed in games over the last 30 years.

The trope of conquest does not merely shape story and graphical representations of place, but also gameplay itself. Longan (2008) discusses the way in which many games have a component of *ordering* a chaotic landscape through both story and play. Taking the *Assassin's Creed* series (2007-, Ubisoft) as an example, players are encouraged to unlock different territories within the game map by climbing to vantage points in order to gain a bird's-eye *view* of the area. This example of the power of the dominant gaze runs alongside a gameplay mechanism that encourages players to pacify regions of the game map through fighting rival powers and tidy the landscape by collecting various items and treasures hidden within it. By the end of the game the player has remade the landscape through conquest until they *own* all that the landscape has to offer.

Methodologically, these qualities give critical scholars some interesting challenges since we become complicit in the power of the gaze as soon as we engage with any game environment, particularly since that environment has been constructed by an industry which has a clear problem with gender. I have written elsewhere (Jones and Osborne 2019 in press) about a project undertaking a landscape analysis of the reconstruction of mid-19th century London presented in *Assassin's Creed: Syndicate* (2015, Ubisoft). Rather than using a traditional discourse analysis approach to the game text, we placed a group of mostly novice players into the game for a fixed period and then asked them to discuss their experiences of the virtual landscape.

Different approaches can be adopted to capturing player interactions with game landscapes. In the *Syndicate* project we adopted what might be described as a 'play along', adapting Kusenbach's (2003) notion of the 'go-along' where researchers accompany participants as they make everyday walks around their neighbourhood. Participants were left alone to play the game for 20 minutes having had a basic briefing in how to walk, run and look around but not how to use the game's more complex parkour functions. This gave novice gamers the chance to experiment with exploring the city without the self-consciousness of having someone else present. Participants then watched back a video of their gameplay while undertaking an unstructured interview with the researchers.

From a methodological point of view, this is quite a useful project to reflect on. When picking a game to use for this kind of intervention, a good deal of care has to be taken to use a game that will actually help answer the research questions (Arnott 2017). This may seem like an obvious point, but games are very rich texts with multiple layers to consider when designing an activity for participants. *Syndicate*, for example, gives the choice of whether to play predominantly as a male or female character. Many male characters in games are depicted as hypermasculine and, as a player, I find the 'cocky' male character trope to be irritating and hence normally play as a woman whenever given the choice. This personal choice also reflects a fairly common phenomenon of players using virtual spaces to try out different gender identities (see, for example, Todd 2012). Because of this, in the project where we used *Syndicate* as a game landscape to navigate, participants were playing a version of the game that I had already completed and they therefore defaulted to playing as the female character.

This had some unintended consequences for the research. The intervention had been designed to create an indoor proxy for urban walking to test a device for measuring physiological response (see Chapter 3). Thus, we chose to use a near completed version of *Syndicate*, with all areas of the city accessible and enemy characters unable to attack unless actively provoked so that participants could wander freely. Participants talked about a range of different issues but, to the surprise of myself and my collaborator, questions around gender came out very strongly in the interview data, with comments about how the main female character would have been perceived and treated had she been walking around the real Victorian London (Figure 4.1).

Given that landscapes are created through the interaction of bodies and environments, it should have come as no surprise that in discussing the game landscape, our participants reflected on the position of women within that

Figure 4.1 The character of Evie Frye from *Assassin's Creed: Syndicate*. The outfits she wears have a distinctly steampunk air and contrast sharply with the conventional period clothes worn by other female characters within the game.
Source: Author's gameplay / Ubisoft.

world. Not having thought about this in advance no doubt reflects my own gender privilege in how I think about navigating both material and virtual environments. It was, of course, naïve not to have thought about the gender of the character when setting up the intervention but, in this case, we were lucky that this ended up being quite productive for the project. What this emphasises, however, is the importance of carefully thinking through how a complex game text might be interpreted by participants and the effect this could have on the overall aims of a project.

Game landscapes do not simply appear on screen, however. Locative games explicitly blur the boundary between virtual and material landscapes. Such games usually rely on the user's mobile phone to derive GPS location, triggering different parts of the game on the device as players move around in the real world. In a sense, locative games draw on a rich psychogeographical (Debord 2006 [1958]) tradition in that they can act as tools to get people to both explore new areas and see familiar neighbourhoods in a new light. The best known of these games is *Pokémon Go* (2016, Niantic) which encourages players to walk around their neighbourhoods collecting creatures that appear on their phone screen as they move into different physical locations. The game created a blizzard of commentary upon its release for its potential to enhance physical exercise, reduce social isolation and improve well-being outcomes among otherwise sedentary children (among many others, see LeBlanc and Chaput 2017, Tateno et al. 2016).

From a research point of view, such games are interesting because they can prompt participants to reflect on the ways in which they engage with their everyday physical landscape surroundings beyond the game. Witkowski (2018a), for example, has undertaken a detailed ethnographic study of *Zombies, Run!* (2012, Six to Start). The game places the player as the survivor of a zombie apocalypse, being given audio prompts to run in particular ways to avoid zombie attack. An 'exergame' designed to be used by joggers, Witkowski notes how it deliberately interrupts participants' running rhythms, demanding occasional increases in speed as part of a 'chase' with the sound of approaching zombies intimidatingly loud in the players' headphones. The fieldwork for this project coincided with a series of sex attacks on runners in Witkowski's home town of Melbourne, however, which was curtailing the freedom of women to engage with public space. *Zombies, Run!* had a large female user-base who, through the app, were exploring their neighbourhoods in new ways. As a result, by examining players' use of the game, Witkowski highlighted ongoing tensions in the gendering of public space.

Thus, games and gaming technologies can be used to examine serious issues around how people relate to landscapes around them. Some of this serious purpose can also be seen in the rise of the so-called 'walking simulators'. These are games which use many of the design cues and effects of conventional games but which create interactions and stories that are explicitly slower and more contemplative. The best known of these is *Dear Esther* (2008, remastered 2012 and 2016, The Chinese Room) which started life as custom mod built in the Source Engine (Pinchbeck 2008). Set on a fictional Hebridean island, players walk along wind-blown cliffs, down to beaches, splashing through rockpools as they head into caves, triggering the recollections of an unseen narrator as they pass into different parts of the game landscape. The developers described it as a ghost story as, through the narrator, we gradually learn more about the death of a woman named Esther in a car accident.

Walking sims trouble conventional split between narrative and ludic qualities. They need to be played to reveal the story, but the gameplay is not really a major part of the experience. Indeed, Muscat et al. (2016) argue that ambiguity is central to the walking simulator experience, with players encouraged to examine details in the landscape which may or may not reveal more of the story, explicitly creating a space for imagination to fill the gaps. This, of course, of course, raises interesting research possibilities for projects that are less interested in the games themselves as participants' reaction to them. The lack of a skill barrier is useful here as it is feasible to ask non-gamers to explore these landscapes without being overwhelmed by the demands of conventional gameplay. Likewise, the fact that many walking simulators take a relatively short time to play – *Dear Esther* can be completed in about an hour – again makes it practical to have larger numbers of non-gaming participants explore this landscape for themselves, opening the door to research that asks non-gamers to reflect on the qualities of virtual landscapes.

Building virtual landscapes

In the immediate aftermath of the fire that partially destroyed the Cathedral of Notre Dame in Paris on 15 April 2019, Ubisoft gave away copies of their game *Assassin's Creed: Unity*. The game contains a meticulous recreation of the Cathedral and the developers claimed they wanted to give people the opportunity to visit the building in its undamaged state as a kind of virtual tourism. Of course, such an act can be dismissed as an attempt to generate some good publicity, particularly as players had set up an account with the company's Uplay store to get the free version of the game. It can thus be seen as an attempt to wean players away from using Steam, which at the time of writing has a near monopoly on PC game sales.

Regardless of the rationale, Ubisoft's action highlights just how much work goes into creating realistic landscapes within digital games. This activity builds on techniques developed in the heritage sector to document different assets. Buildings can be laser scanned or surveyed using photogrammetry, sometimes using drones to capture hidden features and angles, in order to produce a shaded 3D model (Hess et al. 2015, Statham 2018 in press). For game landscapes, developers will often manipulate some of the elements captured through these techniques, editing detail to increase render speed, simplifying textures and sometimes altering elements of the building to facilitate gameplay. *Unity*'s Notre Dame (Figure 4.2), for example, includes the now destroyed spire that did not exist at the time the game was set but that the

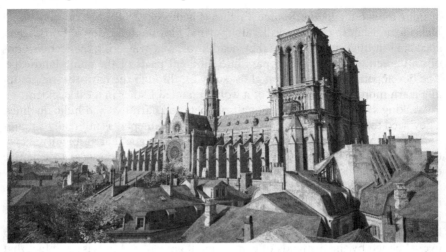

Figure 4.2 Reconstruction of Notre Dame Cathedral from *Assassin's Creed: Unity.*
The game is set around 1789 with the original medieval spire having been
removed three years earlier. The spire depicted in the game was not in fact
built until Viollet-le-Duc's restoration of the cathedral 1845–70.
Source: Author's gameplay / Ubisoft.

developers included because they worried players would be confused if such
an iconic feature of the contemporary building was missing (Webster 2019).

This coming together of techniques developed by the gaming and heritage
sectors has created a significant digital archive of real buildings. With wars in
places like Syria and the Yemen having destroyed many historically important
sites or made them too dangerous to visit, digital recordings may give the
only opportunity to experience them. A small number of these datasets and
models have been made freely available, with others available for a fee. This is
significant because it opens up research possibilities deploying these assets
within game-like environments that the researchers develop for themselves.

Creating a triple-A video game like those in the *Assassin's Creed* series,
takes years, involving the work of thousands of people including programmers, graphic designers, musicians and actors. Smaller games, conversely, may only need a handful of people to produce. *Dear Esther*, could be
produced by a team of less than five people partly because in both original
and remastered states it was built using pre-existing games engines. These
games engines allow smaller developers to get started without having to write
the fundamental code to control elements such as the basics of terrain, gravity
and object collision. Games engines such as Unity and Unreal are available
for free, with developers only being charged where sales of any games built
with the software start to amount to thousands of dollars. As a result, they
have become the default option for small scale development offering tremendously sophisticated tools for zero start-up cost.

The kinds of digital models built for games can be experienced very differently when viewing them using virtual reality (VR) compared to a standard computer monitor, bringing a very acute sense of embodiment to a digital environment. VR has been through multiple iterations over the years although technological limitations and cost has meant mainstream adoption has progressed very slowly (Hillis 1996, Fisher and Unwin 2002). The recent third wave of VR has done much to reduce the cost and produce graphically compelling experiences (creating the 'wow' factor as Heim 2017 puts it). Nonetheless, VR remains expensive despite dramatic reductions in cost and has other significant barriers to wider adoption such as creating motion sickness in a fairly high proportion of users. The promise of VR is that it is a non-mediated experience, with users losing the sense of being connected to the material world because they are fully immersed in the virtual (Bailenson et al. 2004). Indeed, a sense of 'presence' within a virtual environment is a key subjective marker of quality in games (Schumann et al. 2016). But that presence in VR comes with costs as VR cuts users off from the world around them, meaning that they are left somewhat vulnerable, looking slightly ridiculous in a bulky headset with the potential to trip over real objects that do not appear in their virtual world (Figure 4.3). There is also quite a strong sense of *isolation* as the VR user is effectively in a different space to people around them, so that the experience is not really shared.

While VR has significant disadvantages for mainstream audiences, for specific industrial and research uses, it has tremendous value. One of the lesser-known applications of gaming and VR technology is in law enforcement, where laser scanning and photogrammetry are employed to record crime scenes in minute detail so that they can be revisited by investigators throughout a case and potentially used as evidence in court (Buck et al. 2013). Ralf Breker of the Bavarian State criminal office used a games engine to build a reconstruction of the Auschwitz concentration camp, creating a model from a combination of lidar, laser scanning, photogrammetry and archive records (Cieslak 2016). The model can be navigated at an accurate scale in a VR headset and its purpose was to demonstrate what could and could not be seen from different vantage points on the site as part of a criminal case against wartime SS officer Reinhold Hanning. The former guard was subsequently sentenced to five years as an accessory to tens of thousands of murders because the model demonstrated that he could not have been unaware of the murders taking place in the Auschwitz camp at the time he was working there because he would have had a clear view.

Most researchers will not have the time or resources to build anything like as complicated as Breker's Auschwitz model within a games engine. Nonetheless, while accessing the more gamerly functions of Unity requires some programming skill, it is possible to build relatively simple virtual landscapes using existing code 'prefabs' and freely available digital models. This offers some powerful research opportunities, with really interesting work having been undertaken by Bob Stone's lab using this approach to give members of the public a chance to explore hidden and inaccessible heritage sites (Stone

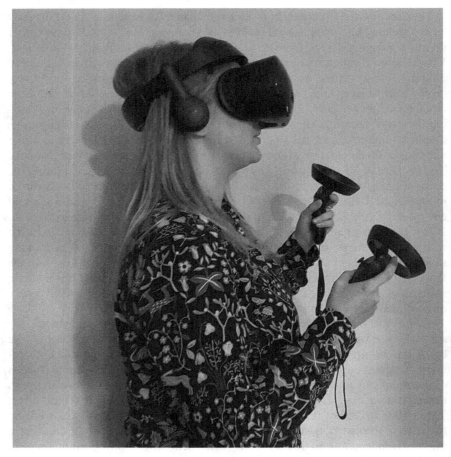

Figure 4.3 A virtual reality head-mounted display. Although incredibly immersive, VR
          headsets are also quite solitary experiences. It can be quite sensible to have
          someone on hand to look after participants using a headset, particularly if
          it is their first time in VR.
Model: Tess Osborne.
Source: Author.

2018). I have done something similar in a less sophisticated manner and there
is something strangely compelling about taking digital models and turning
them into game assets that a virtual player can run around, jump across and
explore (Figure 4.4). Paradoxically, perhaps, being able to *play* these sites
from a first-person perspective makes them feel more *real* than when simply
looking at an overview of a digital model – certainly this approach gives a
compelling sense of scale and immersion in a site which it may not be possible to visit in practice, particularly when viewed through a VR headset.

Figure 4.4 Custom-made virtual environment built using the games engine Unity. This example gives the player a sense of how the University of Birmingham campus would look when walking around it in a scenario where a large-scale windfarm had been installed.
Source: Author.

The sense of presence in spaces that are not accessible is one of the things that has really interested me about working with VR (Portman et al. 2015). There are many reasons why one might not be able to go to a location from cost and safety reasons to the site no longer existing or, indeed, never having been built. My department has for many years been taking students on field-courses to Berlin and, in examining different ideological discourses played out through architecture, I have become interested in the traces of National Socialist urban design that remain in the city. One of the difficulties for students is in conceptualising the scale of the regime's ambition and understanding how the different plans and models would have looked if built. The most significant of these unbuilt designs was for the Germania Welthauptstadt, a version of a neoclassical city which Albert Speer intended to replace much of central Berlin following a German victory in World War II (Scobie 1990).

The plan was ridiculously overblown and probably unbuildable on Berlin's marshy ground. The scale of it is hard to picture; its most significant building, the domed *Volkshalle*, intended to reach an implausible 290m in height. The existing Reichstag building was to have been retained in the design and this gives us useful a point of reference to understand the plans for Germania. I sourced a collection of freely available 3D models of buildings from the

Germania scheme and used Unity to build a VR experience for exploring this unbuilt site. Indeed, we had students using the VR model while standing next to the Reichstag (in both material and virtual form) while in Berlin so that they could begin to get an understanding of just how enormous the planned buildings were.

I have also used this model as part of a workshop in Washington DC organised by my collaborator Tess Osborne along with Danielle Drozdzewski and Jacque Micieli-Voutsinas. Participants stood in the National Mall in Washington DC, giving them the opportunity to contrast Hitler's vision in the VR headset with the material reality of one of the landscapes that inspired his architects. This allowed participants to examine questions of scale and intimidation in the architecture of dictatorship as well as to reflect on the spatial qualities of the 1902 McMillan Plan for Washington DC. Participants were also given the opportunity to try a second VR model, this time a simplified rendering of the National Mall itself, but with a virtual Martin Luther King on the steps of the Lincoln Memorial, with the iconic 1963 'I have a dream' speech playing through the VR headsets (Figure 4.5). This created a prompt to think about how the space would have felt at the time of this important speech in the history of the nation.

Figure 4.5 Users trying a VR reconstruction of Washington DC's National Mall whilst standing in the real landscape.
Source: Author.

Neither of these two VR models can be claimed to be of the highest, most compelling quality. Nonetheless, even relatively poor graphics can be transformed when looked at through VR because of the sense of *presence* that the technology gives in that virtual space. Arguably the experience of building these simple models can be seen as the game-making equivalent of Aarseth's (2003) notion of playing research, meaning that we have to engage with these technologies, even if only at a simple level, in order to understand how they work. This in turn can lead to interesting projects working with designers to explore the working practices of software houses that underpin the construction of their games (Whitson 2018 in press).

Conclusion

The landscapes built within games and the ways in which we interact with them have become increasingly complex and multifaceted. As critical scholars it would be too easy to simply dismiss games from feminist and post-colonial perspectives. The games sector undoubtedly has a sexism problem and gameplay is all-too-frequently predicated on notions of conquest and mastery. Nonetheless, gaming offers very interesting opportunities for research which we can broadly group into three areas with different levels of methodological complexity:

1 projects that build on a cultural studies tradition of discourse analysis extended to include the mechanics of gameplay;
2 projects that employ ethnographic approaches directly engaging with players both in person and via online multiplayer platforms;
3 projects that use games technologies to create environments which can be directly interacted with by participants.

The fact that games are interactive does bring particular challenges to researchers; having some experience of playing the games being studied is an essential first step in any project. Games are more than just flickering images on screens because they blur the boundary between physical and virtual embodiment. This quality of games makes the researcher's embodied positionality a particularly important issue to consider. The games sector can be highly exclusionary both in the (gendered, raced) cultures that shape its communities and in the simple degree of embodied skill required to engage with different games. As a middle-aged person I simply cannot play a game like *Fortnite* (2017, Epic Games) without being instantly killed by players less than half my age whose reaction times are far in excess of my own. As the esports example starkly demonstrates, games research frequently requires a detailed understanding of embodied skill, some of which can only come from the researcher themselves becoming embedded in a gaming community.

Because of the ways that the landscapes of games are increasingly mimicking and intruding into material landscapes, games provide excellent opportunities to create interventions to help understand participants' relationships with their everyday surroundings outside those games. Virtual reality is a really interesting tool in this regard because it can create a total sense of immersion for participants in a very different environment. Research using VR requires compromises, however. VR experiences can be highly isolating and for participants it can be difficult to convey the embodied multisensory experience of being *in* a virtual space. The sense of presence that VR gives, however, can be crucial in training scenarios and projects seeking to expose participants to locations and events that would otherwise be impossible to access.

Unsurprisingly, games technology works well with experimental and playful approaches to gathering research data. Of course, there is always a danger here that we get so caught up in the methodological innovation that we lose sight of the need to generate meaningful research questions. Learning how to build a landscape in a games engine can be tremendously satisfying as a technical challenge but one needs to be careful that this does not simply become a solution in search of a problem. Just because one can build a model of, say, an ancient Egyptian city, it does not automatically follow that there is a meaningful research outcome to such activity. Nonetheless, developing some knowledge about the use of games engines can prove useful in the medium term when seeking to design interventions around a particular research question. Even if the game environment that you need is beyond your own technical capability to construct, by knowing some of the basics of building these environments, it is possible to have a more meaningful and productive dialogue with a technical collaborator – extending Aarseth's 'playing methods' into 'building methods'.

5   Creative practice

Introduction

Richard Long, still a student in 1967, was hitchhiking back to St Martin's School of Art one day when he began repeatedly walking up and down a field in Wiltshire, creating a straight line in the grass with his footsteps. This event now exists only as a photograph, endlessly reproduced, a second-hand reminder of Long's walking act. There is an ambiguity about whether the performance or the photographic documentation is the 'art' in this piece. Nonetheless, the technology of the photograph created a measure of permanence to something ephemeral. Indeed, without the photograph it is unlikely that the piece would ever be remembered or discussed at all.

Art is an inherently embodied practice, but technology and art have always been deeply intertwined. It is tempting to make a crude separation between digital artists and creative practice that employs digital tools. Digital art explicitly seeks to explore how our worlds are reshaped by technology. This is not necessarily the aim of most creative practitioners, yet the majority do use technology to a greater or lesser degree within their work, even if this is just a case of promoting or documenting their practice via social media and online video. As Freeman and Sheller (2015) argue, however, there is not such a clear-cut division between digital art and the use of digital tools, with many artists working at the interface of embodied, 'analogue' practice and more explicitly 'digital' works.

In this chapter, however, I have deliberately chosen to talk about broader *creative practice* rather than *art*. Fundamentally, this allows us to sidestep questions of whether these practices result in something which merits aesthetic judgement – in essence focusing on process and method rather than output. This in turn lowers the barriers to researchers working in this area because it removes the pressure to produce something of aesthetic quality. As a result, there are many interesting opportunities for scholars willing to engage with both creative practice and creative practitioners around the conceptualisation, production and distribution of work through new technologies. Of course, it is easy to dismiss non-arts researchers engaging with creative techniques as mere dilettantish dabbling, resulting in bad poetry and derivative photography. If the purpose is explicitly not to produce work of aesthetic merit but instead to employ different approaches for understanding the world around us, then there can be real value in scholars trying out a range of methods drawing on creative practice.

This chapter breaks into four parts. The first explores examples of how far ideas around the intertwining of bodies and technologies has been pushed by some artists. This gives a framing for the next section examining the nature of collaboration between researchers and creative practitioners, examining power relations and some of the difficulties that arise when attempting to genuinely co-produce work. The chapter then turns to explicitly consider the sensory within technologically mediated creative practice and the place of the non-visual senses. The final section explores some of the practicalities of researchers doing creative practice themselves as a tool within projects. The chapter concludes with a consideration of the opportunities presented by engaging with creative practice in research whilst acknowledging the significant problems with producing meaningful materials from these kinds of research encounters.

Pushing the limits of creative practice, technology and embodiment

Despite having said that this chapter is about creative practice rather than art per se, I am going to turn first to discuss two pioneering artists working at the interface of technology and embodiment who serve as an exemplar of just how far these ideas can be pushed. ORLAN and Stelarc have both separately explored the limits of how bodies engage with technology and have received a great deal of scholarly attention as a result (Goodall 1999, Fleming 2002, Lovelace 1995, Dixon 2019). Rather than focusing primarily on their artistic outputs in themselves, the interest here is in how ORLAN and Stelarc's approach to working with their bodies unsettles our understanding of what research can and should be.

ORLAN was already a well-established performance artist when she undertook her most famous works *The Reincarnation of St ORLAN* (1990–93) and *Omnipresence* (1993). She began using plastic surgery to change parts of her face to match famous works of art: the chin from Venus, the eyes of Diana, the lips of Europa and the brow of the Mona Lisa. Some critics suggested that rather than critiquing the male gaze in art, the surgeries merely reproduced them by transforming her face to meet conventional beauty standards. Such criticism was firmly rebutted when a 1993 surgery resulted in implants being placed beneath the skin of her brow to give the appearance of small horns emerging from her forehead.

Technology is at the heart of the work, most obviously through the surgical interventions themselves but also through the way they were documented. The operations were videoed and broadcast live to a selected audience, with ORLAN herself conscious throughout the surgeries, responding to audience messages sent via telephone and fax. This pioneering interaction with a remote audience has today become routine, banal even, as live streaming with audience feedback is now freely available via platforms such as YouTube, Facebook and Twitch. This highlights a significant change in the distribution of art since the 1990s. While only a select few invitees would ever have seen these films at the time they were being produced, global audiences are now just a click away. Indeed, YouTube has given ORLAN's work a second life, with extracts from the surgical videos now having found their way online, still retaining their power to shock more than a quarter of a century after the original performance.

Australian-born Stelarc's work is more overtly technological, using his body as a canvas for experiments interacting with new technologies. In a deliberately provocative move, he has declared that the body is obsolete. This position opens the door to finding ways of customising our existing bodies to fit new ways of living. His early work was about technological augmentation, moving to Japan in the early 1980s to work with a team of engineers on a prosthetic third arm, which he controlled using muscles in his legs and abdomen. He later undertook experiments wiring his left arm up to electrodes that stimulated involuntary movements in the limb. This led to an early piece of online participatory art where a remote audience could choose to move his body, via electrodes, in ways outside his control.

His 2017 *Stickman* performance used an articulated exoskeleton that forced his body to move through 64 different possible gestures selected via a series of algorithms. As such rather than the technology being used to augment the body's own movements (as was the case with the exoskeleton worn by the character of Ripley in *Aliens*) instead the technology controlled the movement, with Stelarc's body merely a prop upon which the technology acted. This activity was combined with multiple microphones that recorded the sound of his limbs being arbitrarily moved and transformed this into a piece of audio art.

In the cases of Stelarc and ORLAN the body becomes a vehicle for challenging our understanding of what a technologically mediated world *is*. This is not, of course, a call for scholars to begin to experiment with extreme surgery as part of a research process, but rather to highlight the ways in which creative practice can refigure what forms research can take. Certainly, among social scientists there is something comforting about traditional methods – surveys, interviews, etc. – because of their familiarity and robustness. The lesson here is that if we are not willing to explore more uncomfortable (metaphorical and literal) ways of doing research, there is a danger of missing out on significant areas of human experience, particularly as technology continues to drive considerable changes in how we relate to the world around us.

Approaches to collaboration

The engineer Billy Klüver collaborated with the cream of New York's alternative art scene in the 1960s – including household names such as Andy Warhol and Jasper Johns – to help realise their artistic visions. He led a team of 40 engineers who collaborated on the celebrated *9 Evenings* event which combined performance art with novel technology. He subsequently went on to found EAT (Experiments in Art and Technology) which helped bring artists together with engineers to produce new experimental works. In a later interview he talked about his approach being founded in an understanding that 'one-to-one collaboration between two people from different fields always holds the possibility of producing something new and different that neither of them could have done alone' (Miller 1998, 28). He also talked positively about the fact that new technology developed since his 1960s heyday had made it easier for artists to undertake some of this kind of work without external assistance but that at the same time the barriers to collaboration had also been lowered. Today it is no longer particularly unusual for an artist to seek to work with a technical expert to help deliver a project.

Klüver's approach, like that used by ORLAN and Stelarc, was that the engineers, scientists and other experts collaborated with the artist in order to realise the *artistic* vision. Of course, the act of collaboration with a non-artist changes the nature of the work, but the artists were the ones controlling the process. From a researcher's point of view this can be stimulating, but ceding control of the process does not necessarily deliver on your own research agenda as the researcher becomes merely an assistant to someone else's vision. I have been involved in an example of this kind of *art-first* project where myself and a group of other scholars helped the performance artist Emily Warner explore a dataset of geotagged Flickr images. Emily's response to these images was to create a series of performative interventions in spaces where some of the original photographs had been taken. Although we produced a collaborative academic output from this work (Osborne et al. 2019), the main outcome from the project was Emily's artwork, produced as part of *Radical Sabbatical*, an artist in residence scheme at the University of Birmingham.

Of course, artist in residence schemes can provide valuable inspiration on both sides, although as Adler (1979) noted many years ago, simply the fact of working in an academic 'office' environment changes the nature of artistic practice. Residencies can provide interesting opportunities for artist and researcher to learn about each other's practices and expand their professional, methodological toolkits. Foster and Lorimer (2007) have reflected on this kind of evolving artist-academic collaboration, noting, for example, that academics tend to be less adept than artists at working with the visual and in understanding the potential of different performance spaces. Artist-scholar collaborations can be quite appealing simply because they are *different*, hard to categorise and audit by academic managers, with artists lending a certain cachet of the unconventional to academic work.

In any interdisciplinary work there is an emphasis on learning at least some of the disciplinary 'language' used by collaborators, their epistemologies and working practices (Jones and Macdonald 2007). One of the risks for scholars working in these hybrid areas is simply in asking how the outputs should be judged against conventional measures of academic or artistic success (Shanken 2005). Collaborative outputs often carry less disciplinary weight and thus do not always bring career advancement; the practical outcome of this is that work with artists can sometimes be pragmatically reduced to the shiny bauble within a project rather than a core concern. In these circumstances the artist is usually brought in to help with a public engagement strategy, generally by producing something visually 'cool' that helps to sell the research (Kemp 2011). Reviewing funding initiatives for artist-scientist collaborations, Edmonds and Leggett (2010) emphasise the need to set clear outputs for both the artists and the scientists so that both parties can achieve their goals. They also highlight the importance of relationships being allowed to develop over the longer term and note that evaluation of the collaborative process must be built into all stages of any project if that process is to be meaningful.

The difficulties surrounding researcher-artist collaborative work are not only a problem for academics – creative practitioners themselves are concerned about how their work gets entangled in the agendas of other disciplines. We saw this during the UK's New Labour governments of 1997–2010 where a great deal of energy and effort was devoted to promoting the creative industries as being a key generator of economic growth (DCMS 1998, 2010). In this period there was a deliberate obfuscation of the boundary between different kinds of creative labour and artistic production, with profitable industries such as film, TV, music, software and gaming put into the same category as museums, galleries, community arts and other sectors requiring significant public subsidy. Although software was later removed from the government's definition of creative industries, those within the cultural sector continued to emphasise the importance of seeing the profit-making and publicly subsidised elements of the sector as working in tandem. Claims for this interconnection reached their high water mark with The Warwick Commission on the Future of Cultural Value which discussed the 'cultural ecosystem' as a way of emphasising the connections between different elements of the creative and cultural sector (The Warwick Commission 2015). The idea was that, like an ecosystem, the wider creative economy would suffer in unpredictable ways if elements of the arts sector had their public subsidy withdrawn.

The claim that the cultural and creative sectors work as an 'ecosystem' is highly contestable (Jones et al. 2019). Nonetheless, the reason for discussing these changing discourses about the creative economy is that the New Labour period gave rise to a great deal of emphasis within the arts sector of undertaking *socially engaged*, collaborative work with communities. In a highly cited critical review of the 'social turn' in art, Claire Bishop (2005) noted how funding and assessment of quality had been reduced to a series of metrics, prioritising social impact over aesthetic merit. Collaborative working, particularly co-constructing outputs with communities, became a measure of value, with Bishop arguing that:

artists are increasingly judged by their working process – the degree to which they supply good or bad models of collaboration – and criticized for any hint of potential exploitation that fails to "fully" represent their subjects, as if such a thing were possible.

(ibid., 180)

A spin-off of this change in discourse was that, within some disciplines, collaborative work with artists became valued more for its social than its aesthetic potential. One can see this discourse playing out within urban planning, for example, where there has been a trend toward using artists as key actors within placemaking. Artists have been brought in by local authorities nominally to involve communities in public engagement processes around the redevelopment of neighbourhoods. This can have the effect of distancing a local authority from the community being impacted by a redevelopment because the involvement of an artist is seen as ticking the box for public engagement. Such a tick box approach takes little account of whether or not the artist is equipped or inclined to co-create a vision of place with that community (Bain and Landau 2017).

The significance of this for researchers is that the social turn in art has brought a dangerous temptation to use creative practitioners instrumentally as a 'way in' to doing work with community groups. I will hold my hand up to having been guilty of this on occasion in work that I have done when researching questions of creative economy (for a critical account of some elements of this work see Isakjee 2019). It is not unreasonable to wonder whether there might be better ways of engaging with communities than what has now become somewhat of a default action of bringing in an artist to 'facilitate'. This is a particularly acute issue within anthropological and ethnographic work where there is an ethical responsibility on the part of the researcher to work closely with their subject community to address the potential for misrepresentation. As Strohm (2012) reflects, artists do not have the same disciplinary constraints on how communities should be engaged or represented. Thus while working with an artist embedded within a community might open ethnographic fieldwork up to new perspectives and new ways of thinking about a particular group, the artists' approach may work counter to the researcher's disciplinary norms about how those groups should be represented and the ethics of field research (Schneider 2013).

While many artists are professionally committed to doing collaborative work with researchers, industry, communities and others, it is an easy critique to suggest that others are falling back on this kind of activity because they have been unable to make a living through the products of their own aesthetic practice. As scholars we therefore have to guard against using arts collaboration simply as a lazy way of working with communities or generating a pretty output for an end-of-project public engagement activity. Instead we must think carefully about the different perspectives that can be garnered through such a collaboration and how working with a particular creative individual helps us to answer our research questions.

Of course, it is not always necessary for the artist and researcher to agree on all aspects of the project. In a fascinating piece of collaborative writing, Koski & Holst (2017) reflect on a project exploring the increasing trend for vaccine hesitancy among new parents. Vaccine technology is quite literally inserted into the body and particularly since Andrew Wakefield's (Wakefield et al. 1998) thoroughly discredited study that linked vaccine use to autism there has been a significant minority of parents who have refused to vaccinate their children. Holst is a vaccine scientist who collaborated with Koski in order to gain a better understanding of the motivations of parents in this group. In addition to conventional interviewing, Koski responded with a series of visual sketches exploring the narratives produced by the parents about how vaccines interact with the body. What is so fascinating about this study is that Koski shared some of the participants' hesitancy about the use of vaccines, putting her at odds with her scientific collaborator. This fundamental disagreement brings a very personal dynamic to their collaborative paper examining these debates.

Where I have worked with artists, it has tended to be with individuals with whom I have developed a positive personal relationship rather than necessarily because we absolutely agree on what a project is trying to demonstrate. In the past I have collaborated with the spoken-word poet Chris Jam on projects exploring creativity and sense of place. His creative practice is designed to be heard rather than read meaning that audio and video recordings are crucial to disseminating his performances beyond a live audience. Responding to a funding call for creative research activities, he and I co-designed a project collecting scraps of favourite poetry and oral testimony from people in Cardiff with the idea of creating a short film bringing these amateur performances together with Chris's own poetic response to the city. Working through the medium of video fundamentally shaped the project, however, not least because it meant finding people who were willing not only to share their thoughts with Chris, but also to have a camera pointed at them while they did so and sign a release allowing the film to then be shared online. The camera's gaze reinforced the power of the artist-researcher seeking to harvest information from a somewhat reluctant group of potential participants. The camera's gaze also ensured that many other voices were silenced by self-exclusion from the project.

The final film was shot and edited in a single day, ready to be shown at a community research festival the following afternoon. Sitting with Chris that evening as we skimmed through the day's footage, we attempted to assemble something that not only captured something of the atmosphere of place in Cardiff that would not have been so easily explored through conventional means (Jones and Jam 2016), but which also had aesthetic merit, doing justice to Chris's status as a performer. We considered the Cardiff film to be a success in its own terms, adding new perspectives to academic debates around 'atmospheres' and showcasing Chris' work as a performer. A more ambitious follow-up project in the Birmingham neighbourhood of Balsall Heath was, however, significantly more problematic and, frankly, in the end had very little to show for the work undertaken.

In Cardiff a member of the research team acted as camera operator, but in Balsall Heath we were working with a local community member – I'll call him David here – who was an amateur film maker in his own right. This led to tensions between Chris and David over their different visions for the film, not helped by the fact that I was not always available to mediate the discussion between them. We had planned for several days of shooting and editing but once filming was complete, David's progress with editing the film was very slow and I eventually suggested that he pass the footage over to me to do an initial cut. At this point I discovered that much of what he had filmed was unusable, with mistakes in the framing of shots, missing audio and the lens at times partially obscured. Chris had done excellent work in eliciting material from residents of the neighbourhood, some of which we were able to use in a subsequent research paper examining changes to the neighbourhood over time (Jones et al. 2017). At the same time, however, these technical problems meant that the aesthetic potential of the project was severely curtailed.

I salvaged what I could and produced a reasonable short film for a non-academic audience but it was a disappointing experience for both myself and Chris. There was a clear power imbalance between artist and researcher here which contributed to the problems in producing a quality aesthetic product (Kemp 2011). I had involved David in the project in part so that I could say that the work had been *co-constructed* with local community members – a central claim in the kind of research I was undertaking at the time. Given that I was the one who had secured the funding that was paying for Chris's time on the project he was clearly going to be less inclined to go against my wishes that he should work with an amateur film maker in order to meet my research objectives. Chris never criticised me directly for this, though it was clear he was not happy at the working relationship with David and was furious at the technical problems he created. The finished film was eventually shared online but was not what it could have been and although the poem that Chris wrote for it found its way into a subsequent book (Jam 2019), I still feel a little guilty about how things worked out for him with this project as it lacked a significant performance-led output for him.

Creativity, technology and the multisensory

It used to be something of a truism discussed among my colleagues that the least imaginative undergraduate students proposed doing dissertation projects on football. In more recent years we have, instead, been inundated with ideas for dissertations looking at visual representations on Instagram. Fortunately, this is a far richer seam to explore, bringing up an array of diverse issues all the way from body image and fitness (Tiggemann and Zaccardo 2016) through to local politics and journalism (Ekman and Widholm 2017). There are, of course, a great many visual artists who use platforms like Instagram, but most of the posted photographs are aesthetically uninteresting and derivative. Nonetheless, the platform gives users the opportunity to express themselves and, through the use of filters and other editing tools, they are explicitly encouraged to think about their photos as a means of demonstrating their creativity. This said, one should not get too carried away seeing a great democratising of creative production facilitated by these new technologies. As Reeves (2014) has demonstrated through an analysis of the UK government's *Taking Part* surveys, far and away the most significant predictor of engagement with arts and creative practice remains the level of educational attainment.

The popularity of social media platforms like Instagram nonetheless highlights how contemporary society has become saturated with images. Our lives are increasingly driven by the visual register, which raises interesting research questions about the place of the multisensory. The dominance of the visual register in creative practice brings a problem for critical scholars (Pallasmaa 2005). We touched upon Foucault's analysis of the panopticon in Chapter 2 in specific relation to privacy. The panopticon controls its prisoners through the visual – the gaze of the guards being an act of power. Feminist scholars remind us that the act of looking can be an act of domination, particularly in reducing women to objects of consumption for male pleasure (Rose 1992). The tech sector does, however, prioritise visual interactions, in part because these are easier to reproduce digitally than stimulation of the other senses. Arguably, then, the power of the gaze is coded into digital systems, which is a concern given that many of the projects discussed in this chapter rely upon photography and video as a key mode of production and distribution.

The Chatterley trial in 1960 responded to the Obscene Publications Act, 1959 which sought to give greater protection to the status of literature whilst at the same time clarifying the law toward pornography. The ban on DH Lawrence's *Lady Chatterley's Lover* was overturned in part because it was judged that the work as a whole had literary merit beyond its explicit sexual content (Rembar 1969). The distinction between art and pornography is a fraught debate that I have no intention of rehearsing here. Doubtless, however, even if it is not considered to be art, pornography – and sex work more generally – can certainly be thought of as a *creative* practice and one which has been fundamentally shaped through new forms of technology. Video streaming, for example, is now a taken-for-granted part of the Internet. Yet this technology developed rapidly in the late 1990s primarily to serve a vast untapped market for online pornography which placed a premium on high image quality, anonymous consumption and ease of availability (Brooks 1999). In its conventional masculinist forms, pornography, like Instagram, can be thought of as focusing purely on the visual register. This is seen acutely in VR pornography, where male consumers of heterosexual imagery respond negatively to seeing (most) male body parts because this breaks the illusion of the user himself being physically *present* in the scene and thus the sense of immersion in the experience is lost (Rubin 2018). This leads to challenges in how this material is filmed, with male actors trying to keep their hands out of shot, for example, because they do not look like they belong to the body of the consumer. The actors' bodies in these cases are reduced simply to fulfilling the male consumer's fantasy of having above average-sized genitals.

The wider tech sector has a somewhat uncomfortable relationship with the sex tech industry, which is often relegated to basements or side rooms at major sales conventions. We see this discomfort in the controversy over the Osé vibrator. The Consumer Technology Association gave this product an innovation prize in 2019, only to withdraw the award and ban the manufacturer from the subsequent Consumer Electronics Show, before an apology was issued and the prize re-awarded months later (Zraic 2019). This threw into stark relief the double standards of a tech sector that has no apparent problem with using scantily-clad 'booth babes' to attract audiences to convention stands but was uncomfortable about a sex tech product designed by and for women.

Innovations emerging from the sex tech sector are, however, extending online sex work of live and recorded performances beyond the visual register. Teledildonics use a variety of mechanical devices connected via the Internet to transmit and reproduce the haptics of sexual encounter (Liberati 2017). These technologies are in part sold on the prospect of allowing intimacy at a distance – a loved one being away from home for example – as well for the creation of technologically-augmented heightened stimulation (Nixon 2017). They can also be connected to professional sex workers as a virtual extension of their creative practice. This last is significant because it asks us to think about the power relations inherent in any kind of multisensory immersion through the virtual. In essence, it is not simply the gaze that can be construed as an act of power in a technology-mediated world.

From a research point of view, we are therefore reminded to consider these power relations and the position of the body in the different forms of connection facilitated through creative uses of technology. We also have to take account of the fact that, unlike the body in virtual space, the physical body is not a neatly sealed unit – it leaks, perspires and rubs up against the messy chaotic world around it. As a result, there can be some quite banal, practical concerns that significantly shape how a multisensory engagement with the virtual can be realised in a research context. When Valve released the VR Index headset in 2019, for example, its controllers were subject to particular praise for the sensor arrays that allow accurate tracking of finger position for reproduction in a virtual environment. These hard, plastic controllers might at first glance seem like a poor compromise for tracking hand movements compared to a VR 'glove' which more naturally follows the form of our bodies. Gloves are, however, difficult to fit to different shaped bodies, quickly become sweaty and are hard to clean (Robertson 2019). Participants are going to be reluctant to get involved with creative activities within our research projects if this involves being strapped into slightly grimy piece of equipment that has just been unfastened from another body – hence one appeal of hard, wipe-clean plastics in the interface between body and virtual world.

One way around this problem is to have participants use their own equipment in projects. In practice, this generally means using their smartphones and although these devices contain a variety of different sensors detecting movement and possess mechanisms for haptic feedback, most projects using smartphones focus on the visual sense. The use of GPS-enabled smartphones in research will be discussed in more detail in Chapter 6; here, however, I want to reflect on the specific possibilities offered by locative art. These artworks require you to visit a particular geographic location to view them but they are generally invisible to people passing by. The novelist William Gibson, more famous for having invented the term cyberspace, used locative art as a subplot within his novel *Spook Country* (Gibson 2007). One of the characters in the novel creates an artwork which depicts the body of River Phoenix as he lay dead in front of the Viper Room in Los Angeles following an overdose in 1993. In the novel the work could only be viewed through a VR-style headset when stood in front of the actual nightclub on the Sunset Strip.

At the time Gibson wrote the novel, the artwork he described was right on the limits of what was possible with the technology available. In the years since, however, what was science fiction has become a practical reality and it is surprisingly straightforward to create augmented reality (AR) experiences. The most impressive way to achieve this is by using an AR headset such as the Microsoft Hololens, a pair of transparent goggles which project virtual objects into the user's field of view such that they appear to exist in the real world. While this is close to the version of AR that Gibson imagined, the Hololens remains a rather expensive device, generally intended for industrial, medical and military applications. Microsoft in fact faced something of an employee rebellion when it secured a $479m contract to supply the US military with the devices to 'increase lethality' in combat settings (Good 2019). For consumer use, however, the most common form of AR is based on smartphones. A digital object is added as an overlay to the view shot by the phone's camera such that this non-material object is fully integrated into the material world as it appears on the screen (Figure 5.1).

One of the most visible consumer uses of this technology came with the creation of *Minecraft Earth* in 2019. Microsoft employed the same technology it was using to put combat information into the eyeline of US soldiers to allow millions of children to place Minecraft objects into the streets of their neighbourhood which they could 'see' via their smartphone cameras. Minecraft, like Instagram, sells itself on allowing individuals to tap into their creative side and *Minecraft Earth* is in some ways the culmination of a decade of experiments in placing virtual artworks in material contexts.

It is striking when reading pioneering work in this area at just how quickly the technology has advanced. Zoellner et al. (2009) for example, writing just at the start of the smartphone era, needed complex custom algorithms to join up the view from a camera via a computer to a portable screen in order to overlay photographs of historic buildings onto the contemporary landscape.

Figure 5.1 Augmented Reality model seen through a mobile phone camera. The model appears to pop-up on top of a printed map to which it is anchored within the app.
Source: Author.

The same effect can today be reproduced with about 20 minutes' work using the Vuforia plugin and pre-made code packages within the Unity gaming engine to create a simple app that can run on even quite basic smartphones. Indeed, I have played with this technology myself to create small models of development plans that seem to materialise on a table so that groups of people can walk around looking at the same virtual object from different angles through their smartphone screens.

There are a number of examples of creative works using AR technology in an activist mode to flag social injustice embedded into everyday landscapes. This can be something playful such as adding virtual graffiti to a corporate logo – a cracked pipe belching smoke that appeared whenever a smartphone camera recognised the BP symbol for example (Skwarek 2018). Such works highlight just how ubiquitous the branding of these corporations is within our public space and their dominance within the global economy. Of course, despite an initial moment of being impressed by how the technology works and mild amusement at the gentle satire, such AR experiences are hardly going to lead to an overthrow of capitalism, any more than Situationism's playful interventions did in the 1950s (Debord 2006 [1958]). There is also a degree of preaching to the converted in reflecting on who the audience would be that would bother to download the app and seek out locations where they could see a campaigning AR intervention.

AR does, however, allow us to think about the nature of the intersection between material and virtual objects. One is not bound to place an artwork in a specific location but works depicting real locations lose some of their meaning when removed from the actual site where the events being projected occurred – as would have been the case with the fictional River Phoenix installation. Conversely, some works deliberately play with geographic dislocation, such as Holloway-Attaway's (2013) attempt to use AR within a suite of other techniques to relocate Herman Melville's novel *Moby Dick* to the small naval town of Karlskrona in Sweden. Thus, while AR technology itself may be a visual technology, as an *experience* AR is given meanings through the multisensory locale in which it is placed. This can be seen in experiments such as the 'Taking the artwork home' app, which encouraged visitors to place virtual reproductions of artworks into their own homes (Kljun et al. 2018) changing the relationship between public art and domestic space. We also see this in sanctioned and unsanctioned 'hacks' to existing gallery spaces, where AR artists insert digital objects that respond to the works on show. These have become increasingly common since the release of Apple's second generation of ARKit in 2018 which made it even easier to build these experiences for iPhones (Katz 2018).

What is cutting edge within experimental creative practice quickly becomes ubiquitous and clichéd, however. The delighted surprise elicited at simply seeing a digital object appear in a real location will soon give way to a taken-for-granted disinterest, particularly among the generation who played *Minecraft Earth*. Papagiannis (2014) has described AR as a medium in transition, as practitioners examine its *aesthetic* possibilities. Similarly, this is also a good moment for scholars to think about the kinds of *research* questions that can be asked and answered through the creative confrontation of the material and the virtual offered by AR. At the time of writing, however, this remains an under-explored topic among critical scholars.

Doing creative research

For academic researchers in more applied creative fields, there is nothing unusual about undertaking research that is interwoven with creative practice. The music scholar Andrew Kirkman, for example, works primarily on liturgical music from end of the medieval period. One of his more recent projects has been in looking at the production of alabaster carvings in the Nottingham area and the ways in which sculptures with sacred themes and Latin polyphonic music had parallel histories as they circulated in England during the Long 15th Century (Kirkman and Weller 2020 in press). This research led to a series of practical explorations through Kirkman's musical direction of the Binchois Consort, which has recorded a number of critically acclaimed CDs responding to this theme (most recently Binchois Consort and Kirkman 2019).

Few of us could ever aspire to produce anything quite as beautiful as the music that Kirkman records as part of his research practice. This can unfortunately be seen as a major barrier to doing creative work because of the sense of frustration that a lack of embodied skill means any outputs produced will come nowhere close to meeting expectations of aesthetic quality. This was an issue discussed many years ago by the psychoanalyst Marion Milner (writing as Field 1950) in her classic *On not being able to paint*. Milner recounted her frustration at attending painting classes only to learn how to produce 'tolerably good imitations of something else' (p. 1) rather than being able to express ideas within her own voice. Thus, instead of trying to paint in a more technically correct fashion, she decided to go back to basics, attempting to develop her own style to reflect the ideas and feelings she was seeking to evoke – producing work lacking technical aesthetic merit but which captured something of her thought processes.

Community-focused work often asks participants to engage – as amateurs – in different forms of creative practice without expecting the outputs to have aesthetic merit. This work allows researchers to tackle questions about social exclusion, belonging and so forth by giving participants the confidence to express themselves through different media (for example, Wright et al. 2006). Why then should researchers themselves not seek to explore the different perspectives they can generate on a topic through undertaking a creative activity themselves, whether or not that activity would be artistically interesting? If there is an overarching theme for this book, it is in having the confidence to try things out, to learn the basics of an approach, even if that is simply a case of closing off blind alleys or finding a shared language to use when subsequently working with experts in a particular area.

'RIDE' was a performance piece that I undertook in 2009 that involved virtually writing the word 'ride' across part of my local area using a GPS tracklog and a bicycle (Jones 2014). It is not a particularly interesting work aesthetically (Figure 5.2), nor especially original – a number of artists were experimenting with GPS as a means to document their movements as early as the 1990s (O'Rourke 2013, 126). Where RIDE was useful, however, was in disrupting my own understanding of these neighbourhood spaces by forcing me to interact with them differently. Although the streets were familiar, I would not have normally cycled down them as part of my instrumental travel around the city. Cycling a peculiar, somewhat arbitrary route allowed me to think about our everyday engagement with these spaces, how comfortable they are to be in when one is not simply passing through from A-to-B. It also forced me to confront the fact that the city I live in is not very amenable to a playful use of spaces designed for the rapid movement of motor vehicles. Of course, I might *know* that Birmingham is not a cycle-friendly city but this technologically mediated performance confronted me with this reality in a highly embodied manner.

In another example, during the process of producing a paper about rhythmanalysis and the environment (Evans and Jones 2008), I convinced my co-author that we should experiment with turning the paper into a film. This played with notions of rhythm in how academic language sounds when read aloud and experimenting with questions of rhythm in how the film was shot

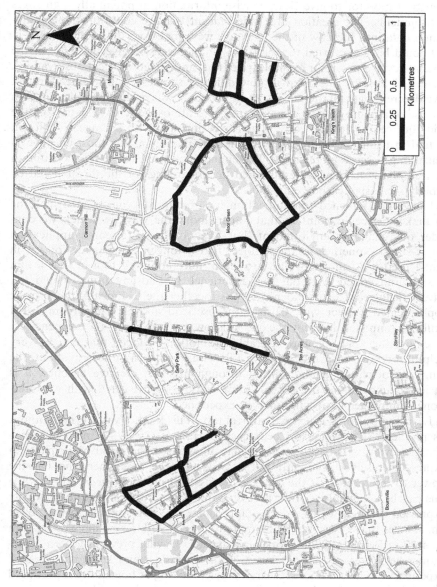

Figure 5.2 Map of a GPS track created while cycling.
Source: Author's analysis. Contains OS data © Crown copyright and database right (2019).

and edited. I was also keen to use the film to experiment with sonification. This is a technique for conveying data through sound (Barrass and Kramer 1999) and is often used in medical applications, for example, the beep of a heartbeat monitor to give clear, concise information in potentially high stress environments. The creation of a video meant that we could integrate sonification data into the flow of the narrative we were conveying in the paper.

I could have converted the data into simple beeps to give the sense of the rhythms over time of, for example, river flow and meteorological observations. Instead, I became interested in how different synthesised instruments might give different musical qualities to these sounds. I was particularly pleased with the use of a piano to give a manic energy to atmospheric CO_2 data from the mid-20th century to the early 21st. I experimented with a range of different virtual instruments before deciding that the piano was the most effective in generating a sense of intensity. The sounds produced felt *urgent*, creating a new emphasis to our well-established understandings of the threat posed by rapidly increasing carbon emissions.

I have not returned to sonification in subsequent projects; it helped me to make a specific point in that paper about the ways in which we consider our changing environment, but I've not had a research application for it since. Nonetheless, that project had a longer-term legacy for me in that I used it as an opportunity to learn more about rhythm in how a film is shot and edited. Without this experience I almost certainly would not have had the skill or confidence to put together the technically much more sophisticated project with Chris Jam described above. While I am not a skilled filmmaker by any means, I know enough now to inexpensively assemble the kinds of unremarkable promotional videos that can be very helpful on any research project. Likewise, by learning more about the language of shooting and editing through my own creative practice, it is easier to understand the potential for work co-constructed with a professional filmmaker than without that experience.

Learning by doing is a very powerful tool. This kind of work is not without its problems, however. As Coemans and Hannes (2017, 42) note in a systematic review, arts-based methods may allow researchers and participants to express themselves differently but it is usually the researcher who has the power to interpret what the creative process and products *mean*. Thus, while this can be a valuable tool for accessing information that might not be easily expressed through words, creative approaches are not immune to the power imbalances that we see in all research projects. A striking example of this can be seen in Pinney's (2005) work on striptease. Having previously been a performer herself, she could act somewhat as an insider and relate participants' experiences to her own in a research project looking at how desire is produced within commercial male striptease. She recalls an incident where one of the workers took her recording device from her to speak into it privately out of earshot. This could be taken as using flipping the power of the researcher, turning the technology against her, but instead she asked herself:

Am I giving, *them*, the Other, power, (as if it's mine to give) or are they doing my homework for me?

(ibid., 721)

The agency of researcher and participant was troubled further by Pinney in reflections on how she paid to receive dances from her subjects, where the charged atmosphere of the club created a sexualised encounter. Clearly the point of these dance performances is precisely to generate that eroticism and the only way to fully engage with this creative practice is to be immersed within it but from an ethical point of view this is clearly troubling.

This example demonstrates that the whole point of creative practice is that *something* is brought into being – whether that be a material object or an atmosphere or a range of other experiences, regardless of there being any aesthetic merit. Trevor Paglen (2009) has talked about this in his work on 'experimental geographies' where he draws upon Henri Lefebvre's ideas around the construction of space. As researchers we inevitably alter that which we are researching, if only in small ways. Action research acknowledges this, shifting from a position where we passively examine the world as it is, toward actively attempting to change things. This is a normal approach in more practical disciplines such as architecture, which creates designs of both material objects and processes, some of which are subsequently brought into being. We see this in the French school of ambiances which seeks to alter the atmosphere of different places (Tixier 2016). A key method employed by ambiance researchers is the urban transect, which involves passing through urban spaces, recording the designs, the noise, light and so forth in order to create a record of how it feels to be in there. The point is in using the material gathered to subsequently create interventions which could resolve particular design issues in those locations, changing the way that it feels to be there.

Although architecture is a practice-led discipline, this more interventionist approach can find its way into other areas of research. One interesting approach to doing this has been the hackathon. The term originated in 1999 and it is usually seen as an event where people to come together to brainstorm and offer solutions to a given problem (Briscoe and Mulligan 2014). Hackathons have their origins in the tech sector and there is often an emphasis on coding, although the core idea of assembling multidisciplinary teams to rapidly prototype responses to an issue has much wider potential. As a researcher, getting involved with or leading hackathons can be a productive means of doing early-stage research and exploring the direction to take with future projects. They can also be very useful as part of postgraduate training to develop skills in working across disciplinary divides.

Hackathons are not unproblematic, however. The tech sector's problems with race and gender have also been played out in how the hackathon model has evolved. The figure of the hacker is almost always represented as masculine, something that was highlighted in an exchange from the first *Matrix* film (Wachowski Siblings 1999) where the hero discovers that the woman he is talking to is the notorious hacker Trinity:

NEO: I just thought ... you were a guy.
TRINITY: Most guys do.

As Brooke (2018) has highlighted, gender cultures continue to be a concern, noting that a celebratory account of hacking's history published in 2010 listed 52 men, 10 computers and 3 women in its who's-who of hacking with all three women being wives of male hackers. As a result, there have been concerted efforts to move hackathons past the classic model of late-night beer and pizza affairs, to create more gender neutral and female-friendly events. 'Hackermoms' for example, was founded in 2012, providing women-only events that include childcare and family-friendly working hours.

Hackermoms and other groups demonstrate that the hackerthon itself can be hacked to mitigate some of the elements of misogyny which are woven through the tech sector. Although these events traditionally have a code-facing element, there is nothing to stop us as researchers using the model for other purposes (Calco and Veeck 2015). The practical emphasis on finding routes to solve an identified problem does differentiate the hackathon from more conventional academic activities such as workshops. The emphasis on creating a solution means that these gatherings become more than simply the kind of unfocused talking shops which are depressingly familiar to most people working within the social sciences. Of course, not every gathering of scholars needs to be fixated on creating outputs, but the hackathon model is a useful tool when more practical solutions are the aim.

Conclusion

Richard Florida (2002) included university researchers as members of his 'creative class', signalling the fact that all academics are creative practitioners to some extent. There is something particular, however, about engaging with creative practices beyond our familiar scholarly endeavours of writing and teaching. Creative practitioners, including those who would self-identify as artists, have developed an interesting range of approaches to understanding the world that can influence the way in which we research. Although many of these approaches require a good deal of embodied skill in order to produce an aesthetically valuable output, that is not a reason for scholars to ignore the potential of these processes for helping us to gain different perspectives on the world around us.

We can characterise approaches to creative practice in research discussed in this chapter as taking three forms:

1   working with an artist to help them to realise their aesthetic vision through techniques that draw on your existing expertise;
2   inviting an artist to collaborate on a research project with different levels of involvement from simple community engagement to full co-construction;

3 researchers using their own creative practice as a tool for developing different insights about the world or as a pilot for future collaborations with creative practitioners.

This is not a clarion call for researchers to churn out terrible video art in the name of research. As with all research approaches, engagements with creative practice need to be undertaken with a view to shedding new light on a set of research questions. Trying out a new practice for yourself can be a good starting point to understand it. Given how rapidly the tech sector is evolving, directly engaging with a process can be incredibly important although, as the striptease example discussed above demonstrates, it is important not to lose sight of issues around ethics and power relations when we do so. Nonetheless, if nothing else, simply trying out a creative practice is an excellent way of exploring what is and is not possible with a given technique to form the basis of a more informed conversation with a skilled practitioner with whom you may choose to collaborate.

Working closely with a skilled creative practitioner is a tried and tested means of generating new research insights. Again, however, it is imperative to think through what it is we want to achieve by working with such an individual. A purely utilitarian approach to these collaborations would be that they have value because they help the *researcher* see the world differently or because they can *sell* that research to a wider audience. In some cases, this might actually be appropriate – not all collaborations need to be equal and there is certainly nothing wrong with enhancing the impact potential of research through an eye-catching end-of-project exhibition. Likewise, academics occasionally help to provide the raw material or expertise from which an artist might produce an interesting aesthetic output with little or no direct gain to the researcher themselves.

Unlike Klüver's model of engineer-artist collaboration, it is not always the researcher who brings specific technical expertise to a joint project with a creative practitioner. Either way, the novelty of this kind of work is not necessarily about simply employing a particular piece of technology but rather in the kind of research questions you are able to answer using the approaches that the technology enables. A major consideration that I have emphasised here is the way that digital practice is heavily skewed toward the visual register because of the way that the virtual is skewed toward this domain. Digital video and photography open up very interesting possibilities in terms of producing and documenting creative practice but also reproduce the power differentials that we associate with the gaze. Again, however, as I have demonstrated here, even the non-visual senses can be interwoven with difficult questions around power in the ways that they are reproduced and stimulated within the virtual.

Engaging with creative practices and practitioners allows us to develop new takes on familiar topics and crucial insights into emerging areas of research. It can be rewarding and frustrating in equal measure but from a researcher's point of view, the stakes are often quite low – if something does not work then all that has been lost is time. It is worth remembering, however, that many creative practitioners are freelancers, thus from an ethical point of view it is important not to waste *their* time. This means thinking carefully about the resources available to pay practitioners, particularly where the outcomes of a collaboration will do more for your career than theirs.

6 Maps, apps and mobilities

Introduction

When I was doing my PhD at the start of the noughties, an Icelandic friend working in urban morphology told me that I would love GIS, the software for mapping and spatial data analysis. A couple of years later when I found time to teach myself the basics, I discovered that she was absolutely right. There has been some fantastic research over the years by critical scholars using and often subverting GIS to take mapping away from its militaristic origins of domination and control. This being said, one of the curious things about working as a geographer, particularly a UK-based cultural geographer, is how unusual it is to be able to use mapping software.

This book has been written to appeal to a readership beyond geography, but this chapter is unapologetically focused on the role of space in how we understand and research embodiment. Two factors help non-geographers here. The first is that the humanities and wider social sciences have undergone a 'spatial turn', acknowledging the idea that 'space' plays a critical role in shaping how we understand and interact with the world around us (for example, Arias 2010, Gulson and Symes 2007, Sheller 2017). Secondly, there has been a shift in the technology of geospatial location which means that it is no longer the preserve of those of us with access to specialist mapping software and technology.

Mobilities has emerged since the early noughties as an interdisciplinary field of study dedicated to understanding a world which seems to be continuously on the move. Work in this area is particularly well-suited to exploring the details of everyday embodied interactions from how we undertake the daily commute, through new working practices, to questions around creativity and leisure. Its diversity as a subject area leaves it open to criticism that it is about both everything and nothing. As a theoretical frame it is valuable, however, because it asks us to pay attention to how movement (or its lack) shapes social processes.

Methodologically, work in mobilities leans heavily on different forms of (auto) ethnography and the use of video. I would argue that mapping and mapping technologies are a somewhat underappreciated area within mobilities research.

Maps must not be treated uncritically, however. When desktop GIS became much more accessible to geographers in the 1990s, there was a lot of concern among critical scholars that they reduced people and places to simple XY coordinates on a screen and thus failed to reveal the complexities of lived experience. Over the last three decades a rich tradition of scholars working in critical GIS have challenged the top-down power structures associated with conventional mapping, working with mapping technologies to empower communities and find novel ways of representing people and place.

The place of maps and spatial data in critical research was turbocharged from around 2006–2010 by the rise of web-based mapping services from companies such as Google combined with the creation of smartphones that included GPS satellite navigation technology. Today a great many of our everyday activities are underpinned by location-based services powering everything from dating apps to house buying. These systems operate almost invisibly, to the point where everyone carrying a smartphone has become a kind of unconscious cartographer. For researchers, these location-aware technologies offer unparalleled opportunities for creating projects where we can trace the movement of individual bodies although not without raising practical and ethical concerns about how we secure the privacy of our participants.

This chapter opens by examining the rise of mobilities as a field of research and some of the questions it allows researchers to ask about embodiment. It then turns to discuss the specific methodologies that have become associated with mobilities research and the possibilities offered by these techniques. The chapter then turns to questions around mapping and the challenges this technology poses to critical research. Finally, the rise of GPS is examined along with the methodological potential of work that uses locative technology.

The rise of mobilities

Thomas Kuhn (Kuhn 1970, 85) talked about paradigm shifts occurring where '... the emergence of a new theory breaks with one tradition of scientific practice and introduces a new one conducted under different rules'. The classic example of this came in the early 20th century where quantum mechanics displaced Newtonian physics in providing a completely new explanation of how light functions. It was therefore with a certain knowing cheekiness that Sheller and Urry (2006) called their much cited editorial introduction to a special issue of a journal 'The new mobilities paradigm'. Describing their ideas on mobilities as a 'paradigm' gives a sense of the authors' ambition to completely change our understanding of the world by examining it through movement.

Emerging from a set of ideas used to examine how the private car refigured cities (for example, Sheller and Urry 2000), Sheller and Urry's notion of mobilities has created a whole subfield of work. Essentially, they argued that the social sciences and humanities had previously tended to think about stability and being *in* place as normal, meaning that lives and things that are constantly in motion were seen as abnormal and even deviant. Instead, they argued, mobility rather than rootedness was becoming the norm for our increasingly globalised world. Of course, there had been plenty of work discussing aspects of our mobile world prior to the development of 'mobilities' as a concept, but it serves as a useful framework through which to understand how movement and stillness shape everyday life.

Mobilities scholarship brings together lots of different ideas and topics to build this argument that movement is the defining characteristic of the 21st century. As such, it can come across as a bit eclectic, grabbing shiny ideas from across the humanities and social sciences. Everything from the transatlantic slave trade through the daily commute and much in between has been considered through this lens. Some of this work explores interesting, if relatively conventional social science issues but with a mobilities twist. Andreotti et al. (2013), for example, examine questions around how neighbourhoods change. They contrast an increasingly mobile managerial class with more rooted local communities and explore the tensions that this can bring to neighbourhood life in different European cities. Here, 'mobility' brings a new dimension to questions of how social class and power protect the interests of wealthy residents.

In terms of this book's focus, however, the emergence of the mobilities debate is interesting because it coincided with a point in the early noughties where mobile technology of different kinds was becoming both ubiquitous and increasingly sophisticated. Indeed, examinations of mobile technologies played a significant role in early mobilities work. The *Travel time use in the information age* project, for example, was led out of John Urry's Centre for Mobilities Research based at the University of Lancaster. Funded by the UK's Engineering and Physical Sciences Research Council, the project investigated how people made use of their time when travelling on different forms of transport, including trains, buses and coaches.

Conventional transport studies suggest that travelling between locations represents unproductive time which should therefore be minimised to add economic value. Lyons and Urry (2005) demonstrated that, conversely, users found a great deal of value in their time spent travelling. Even just the act of staring out of a window or having time to decompress between home and work life could be seen as useful or important by passengers on public transport. At the same time, business travellers in the early noughties were shifting their patterns of work, finding new ways to use the time they spent commuting and travelling between meetings. Given the growing numbers of people working in the knowledge sector, travel time was seen as an opportunity to catch up on different activities, with laptops and mobile phones increasingly being used to facilitate work on the move (Holley et al. 2008).

Since these early studies, the technologies being carried by people as they travel have significantly advanced in portability and sophistication. Rather than it only being the business classes who brought laptops, today travellers without a phone or tablet to use during the journey have become the exception rather than the norm. The new forms of connectivity that are now available have transformed the experience of being away from the fixed places of home or work. Germann Molz and Paris (2015), for example, have demonstrated this by examining recent practices of extended leisure travel overseas. Previously, travellers maintained contact with friends and family back home through occasional letters or expensive phone calls. Even in the early noughties, mobilities researchers found it difficult to maintain research connections with backpackers as they would drop in and out of contact. With smartphones, Wi-Fi and roaming data this situation has changed. Travellers today are constantly connected, keeping blogs, communicating through social media and smartphone photography to the point where:

> Travel mottos like 'I was here!' or 'Wish you were here!' seem obsolete in an age of 'I am here right now and you are (virtually) here with me!'
> (Germann Molz and Paris 2015, 180)

As a result, travelling overseas can be seen as just another activity which is anchored in the everyday routines of home. For researchers this is a boon as instant connectivity and practices of recording travel experiences through social media have made it much easier to examine changing cultures of the leisure travelling experience.

Extended leisure travelling is broadly the preserve of privileged young Westerners, but many others beyond this group have seen the transformation of their travel behaviours through mobile technologies. Porter et al. (2018) for example have undertaken comparative research looking at mobile phone use among young people in sub-Saharan Africa, across two study cohorts 2007–8 and 2013–14. Noting a near doubling in rates of ownership between the two phases of the study, they reflect on the different ways in which phones have reshaped the mobility of young people. One example of this is in students being saved journeys to collect assessment results because of widespread use of text messaging by schools. Many students in the study found this attractive not simply because of convenience but because they were routinely exposed to the risk of violent attack when travelling to and from school. Physical mobilities were thus substituted for virtual mobilities wherever possible.

Writing a book like this it is easy to get drawn into the dystopian possibilities brought by new technologies and forget how they can make positively transformative changes in the lives of people living in very difficult circumstances. Indeed, there can be a somewhat moralistic tone to discussions of technology as mediating different kinds of mobility. Are the travellers studied by Germann Molz and Paris having a less *meaningful* experience compared to those from the 1990s because of being connected in real time to friends and family back home? Is an 'unplugged' travelling experience somehow ethically superior? We see similar moralistic undertones in Vergunst's (2011) study of Scottish hillwalkers. Navigation using GPS-based devices was seen by many walkers in the study as being inherently poor practice compared to using conventional paper maps and reading the terrain, putting inexperienced walkers in danger. There has never been a shortage of people getting into trouble walking in mountainous environments because they were not properly prepared and equipped, however, and the availability of navigation technology is just another factor in a more general pattern of stupidity among some walkers. What we see in the responses by Vergunst's participants, then, is a degree of snobbishness that using digital technology means people are not walking 'properly' (echoing the contempt for 'casual' videogamers among more serious players discussed in Chapter 4). Indeed Vergunst (2011, 213) himself suggests that walkers' navigation 'instruments should allow them to enter into a closer and more direct relationship with the places they move through' suggesting that navigation technology creates an inherently inferior walking experience.

For researchers, the important point here is to resist the temptation to see mobility that is not technologically mediated as being somehow morally superior, more real, grounded and meaningful. As always, it depends on how a technologically mediated experience is framed. A nice example of this is in Holton's (2019) work creating an app ('PlymTour') helping first year undergraduate students explore the new city they have moved to. The app gave students information about the different sites they were visiting, layering additional understandings of the different locales on top of their own embodied experiences. This meant that the perception of an area could be shifted by finding out, for example, that they were standing in what had been the red light district.

Both the content and functionality of the app thus played a role in remaking the city for the students moving around it. This brings us back to the recurring theme in this book, that technologies are not neutral – they are informed by the social norms, preconceptions and biases of the people who develop them. Indeed, as Cresswell (2010) highlights, politics and power relations underpin mobility. He reminds researchers to ask questions about the nature of the movement they are studying: why it is taking place; what constrains or channels it; and what stops it altogether. We must therefore apply Cresswell's concerns about power and mobilities to the technologies that are entangled with the practices of being mobile.

The most visible example of power in mobilities comes when considering the monitoring of those caught up in the criminal justice and asylum systems. Probation and immigration services today often use tagging as a means of recording the movements of people they are monitoring. Tags usually take the form of an ankle cuff, modern versions of which can monitor movements in real time, relaying GPS data back to a control room via mobile phone technology. Prior to these systems being widely employed, individuals being monitored would be 'invisible' for large parts of the day where their presence was not being logged through face-to-face contact (e.g. with a probation officer). Tagging means that they can be prevented from visiting certain spaces and suspect mobilities queried under threat of detention. Thus the technology enables carceral spaces to be extended beyond the walls of the prison (Nellis 2009).

Being able to control movement is crucial to state security but operates in tension with the frictionless borders which are at the heart of neoliberal capitalism. The EU's Schengen free-travel area can be seen as removing the barriers to trade and economic development. Glouftsios (2018), conversely, argues that monitoring at the border of the Schengen area frames the movement of people as a threat, with any individual potentially suspect. A host of technologies, including biometric monitoring, are used to secure those borders against threatening movements. Unsurprisingly, therefore, the airport has been a crucial site for mobilities research, with Schiphol forming a significant case study in Cresswell's (2006) *On the move*. At every level airports use different technologies to sort and filter bodies and their associated luggage, from the automated number plate recognition that identifies suspect vehicles approaching the site, to the databases that identify travellers for additional screening, through various scans and checks to make sure that everyone is in the right place at the right time and are only carrying approved materials. Admittedly, the travelling public occasionally rebels against this, with body-scanning devices having to be redesigned so that operators were not able to directly see the ghostly naked images of people passing through them (Ahlers 2013). Nonetheless, airport security has become a set of rituals, making highly visible the idea that movement can be threatening and must be controlled (Amoore and Hall 2010).

We saw this tension between movement and security being played out in the debates over the terms of the UK's withdrawal from the European Union. Being able to diverge from EU regulations was a core demand of some favouring withdrawal but this came with a requirement to meet stringent customs checks to prevent the movement of illegal products into EU territory. Given that an entirely porous border between Northern Ireland and the Republic of Ireland was a key requirement for peace in that region, a significant part of the Brexit debate revolved around how to square the circle of customs checks with no border stops (McCall 2018). UK politicians called for 'technological solutions' so that the checks could be carried out invisibly, but it was clear that such technology did not exist and was years away from being developed, if it was possible at all (Beall 2018).

This debate brought into sharp focus the temptation toward assuming that technology can provide perfect solutions to intractable social problems, particularly in questions involving mobility. This has been a problematic issue in law enforcement. As we discussed in Chapter 2, facial recognition technology has been trialled by police forces as a means of automating the identification of criminals. Among many other problems, this technology reflects the biases of its creators, being particularly poor at distinguishing between the faces of black people. Police forces have been using technology in less obvious but equally problematic ways for many years, particularly in terms of monitoring and restricting the mobility of ethnic minority populations. So-called 'predictive' policing uses historic crime data to prioritise the allocation of resources to particular neighbourhoods. Given historic institutional racism of policing practices, these approaches build their predictions on the racist assumptions of past offi-cers, reinforcing social division. Kaufman (2016), for example, discusses the creation of 'Impact Zones' in New York city, where areas identified as being high crime have been targeted for intensive, militarised policing. These areas are pre-dominantly non-white and the police patrol in high numbers with an emphasis on street-stops and ticketing even for very minor infractions such as riding bikes on the pavement. This situation has become more problematic as the police are being provided with handheld devices that allow biometric monitoring:

> While all New York police may have fingerprint scanners, the augmented police presence and stop-and-frisk tactics in Impact Zones - even without accounting for officers' racial bias - ensure that biometric data collection disproportionately affects poor communities of color.
>
> (Kaufman 2016, 77)

To avoid the humiliation of being treated as a criminal while going about their everyday business, some of Kaufman's interviewees were more careful about the kinds of journeys they made in their local area. These policies can be seen as deliberately restricting the mobility of people in those neighbourhoods by defin-ing presence in public space as suspect behaviour that can be challenged with the same kind of biometric monitoring that might be seen in the prison system.

This is far from being an issue solely for US police forces. One of my former PhD students, Arshad Isakjee, examined the treatment of young Muslim men in Birmingham for his doctoral research. After the events of 9/11 and the UK's 7/7 terrorist bombings, young Muslim men were targeted for increased attention by the police and security services in the UK. In 2007–8 the West Midlands Police and Birmingham City Council took an opportunity to access central govern-ment anti-terrorism funding in order to pay for a new CCTV camera scheme in two relatively small parts of the city. Both neighbourhoods had a large Muslim population and some 216 cameras were deployed, including automated number plate recognition (ANPR). Some of the cameras were hidden such that people in the neighbourhood had no idea where they were located and thus when and where they were being watched. The mix of ANPR and covert cameras increased the sense that the population of these neighbourhoods had been deemed dan-gerous, with their movements needing to be monitored for the safety of wider society. As the local population became aware of the scheme and the fact that it had received anti-terrorism funding, so a protest movement evolved and even-tually the cameras were removed from one of the two neighbourhoods, but not before significant damage had been done to the sense of trust between the police and communities involved (Isakjee and Allen 2013).

Mobilities and methods

Given that mobilities scholarship markets itself as representing a paradigm shift, it is perhaps unsurprising that there has been significant attention paid to finding new methodologies for examining our mobile world. Büscher et al.'s (2011) edited collection *Mobile methods* showcased some of these approaches and there has been considerable innovation since. Before exploring the diversity of methods, it is important to state that this field has been dominated by variations on ethnography (often autoethnography) and video-based approaches. This is not meant as a criticism; the broad theme for this book is that it is important to find the most appropriate technique for gathering the data you need to answer your research questions. Often, the most appropriate technique may not be the most innovative or exciting.

For investigating the embodied *experience* of a particular mode of mobility, (auto)ethnography can be an incredibly powerful tool and it is unsurprising that it is so commonly employed in this field. In the *Travel time* project described above, for example, mobile ethnographies were undertaken by researchers simply making notes while travelling on major public transport routes. This approach gave detailed insights into the ways passengers used the different spaces of trains and buses, even down to the scale of how they packed and unpacked their bags and staked claims to the area around them (Watts and Urry 2008). Perhaps a more interesting example of (auto)ethnography in mobilities research is David Bissell's (2018) *Transit life*. Bissell is a lyrical writer, and *Transit life* beautifully weaves together accounts of his own experiences with those of his interviewees to create a rich, nuanced examination of commuting practices using Sydney as his experimental site. Done well, as is the case with Bissell's work, this approach can generate novel understandings. It can, however, also lead to rather parochial accounts which attempt to hide underwhelming insights behind claims of providing 'thick' description of an individual mobility experience. The slew of articles about personal mobilities being interrupted by the 2010 Icelandic ash cloud, for example, often seemed more than a little self-indulgent. Indeed, several of these pieces were prompted primarily by the fact that a large number of geography researchers were attending a conference in the US at the time and were briefly stranded (myself included, as it happens, but I never felt the need to write a paper about the experience).

The use of video cameras is a much more overtly technological approach to examining different aspects of mobility. Cameras have become smaller, lighter, higher resolution and substantially cheaper since the early noughties, and so it is understandable that they have been applied to a variety of research projects interested in the movement of people and things. Laurier and Philo were using video on a couple of projects just at the point where cameras were becoming more discreet and affordable. Their work on sales reps used the same kind of hyper detailed approach that would later be used on Urry's *Travel time* project, recording minute details about how commercial travellers went about their daily business (Laurier and Philo 2003). Their subsequent project on cafés used fixed cameras to monitor patrons' comings and goings, generating a huge amount of footage from which they were able to examine the different ways that people engaged with these spaces (Laurier and Philo 2006). Although neither project was overtly framed in terms of mobilities, both demonstrated the potential of using cameras to capture the micro-interactions of people in their everyday use of space. In this regard Laurier and Philo were following in the footsteps of William Whyte (1980) who undertook pioneering work using film to examine how people interacted with different urban spaces, often mounting his cameras on high buildings to get an overview of individual movements.

To accompany his book, Whyte also produced a short film using some of the footage that had been shot for the research. The moving image has an obvious attraction to those interested in mobile life and there have been a number of documentary-type projects of varying quality by researchers in the field. One of the best recent examples is Vannini's (2016) *Low and slow* a 26 minute documentary which aired on Canada's Knowledge Network. In a comment piece about the documentary, Vannini (2017) reflected on how practice-based film-making could enhance mobilities research and bring its findings to a wider audience. Although it has become easier and cheaper than ever for researchers to produce documentary films of a *reasonable* quality, as discussed in Chapter 5, it is important to think about who the audience might be for a filmed output. Indeed, while scholars like Vannini and Laurier are skilled filmmakers in their own right, researchers need to consider whether they have sufficient technical expertise to put together something that an audience might actually want to watch. If not, then we need to think carefully about whether we are filming purely as a means of data collection or whether we want to produce a public-facing output that will require a skilled collaborator to produce.

As video cameras were becoming smaller and better in the early noughties, so it became possible to get a participant's eye view through the use of body-worn cameras. Justin Spinney has been using mobile cameras for many years to produce video ethnographies, particularly of cycling, with some of his earliest fieldwork of this kind dating back to 2004 (Spinney 2011). Likewise, Brown et al.'s (2008) study of off-road leisure cyclists showed the potential for this kind of approach. Using a head-mounted bullet camera, the project captured the lived experience of this kind of cycling which otherwise would be difficult or dangerous to record if participants were distracted by holding a conventional camera. Being able to record embodied mobility practices allows the researcher to understand how expertise is developed and deployed. Their study was also interesting because it was an early consideration of the ethics of using this kind of camera. They noted that presenting the view *from* rather than view *of* the participant not only enhances anonymity, it also gives more sense that the participant is actively involved with co-creating the research rather than simply being a passive object being observed by the research team.

Most of the footage that was captured in these pioneering projects can only be considered useable as a source of data, however, rather than as the raw material for a public-facing output. Early mobile cameras were low resolution, with no image stabilisation producing footage that was nauseatingly shaky. Over the last decade video devices have become ever more sophisticated, with the GoPro in particular now having become a ubiquitous tool for field researchers. These cameras now produce very high resolution, broadcast quality material. Even large budget TV shows such as *Top Gear* and *Comedians in Cars Getting Coffee* now routinely rely on these kinds of devices to capture mobile action in spaces that simply could not accommodate conventional cameras. This quality of recordings can, however, actually bring some ethical challenges around anonymisation as unedited footage of outdoor spaces captures a large amount of incidental detail of people and places that can be easily identified.

One response to this – and to the challenge of representing visual material in published outputs – can be to find ways of remixing still images taken from video footage. Lloyd (2019) has undertaken a really interesting study taking advantage of high-resolution video footage to painstakingly examine how pedestrians look and do not look at cyclists as they pass-by in a shared cycle space. In order to represent these visual cues within the final publication, Lloyd extracted stills from key moments in the footage and put them through a filter to give a low resolution, abstract feel which effectively removed the possibility of identifying individuals from the images. He then presented these video stills in a comic strip style to form a kind of visual vignette depicting different events within the ride. Although I do not personally feel the treated images work particularly well aesthetically (blurring key elements might have been more effective), the approach is an interesting one and highlights the importance of identifying key moments in the footage as the basis for the analysis.

Manually coding events from hours of video footage is, however, tremendously labour intensive. Of course, if one is interested in examining very small moments of interaction, this may be the only means to analyse these data. Very low-cost and high-capacity storage cards bring the temptation to simply leave cameras running but this can create problems later on. The intensity of manual analysis in terms of time and effort mean that it is sensible to plan how footage will be recorded, capturing only what is necessary in order to keep research costs manageable. There are, however, more automated approaches to analysing video footage drawing on machine vision techniques which may be appropriate to explore in some cases. Brandajs and Russo (2019 in press), for example, examining cruise tourism in Barcelona used a mixed approach combining conventional interviews and observations with a quantified analysis drawing on video footage.

Mounting a single, fixed position camera in two public squares, the *Plaça Reial* and the *Plaça dels Àngels*, footage was collected over several days chosen to coincide with cruise ships docking in the city. Using a range of computer vision techniques, the footage was categorised at the pixel level such that pedestrian movements could be automatically detected, thus allowing the researchers to calculate things like how quickly people moved, how clusters of people formed and how long groups stayed within the space. They ground-truthed the automated analysis through site observations such that they were able to calculate how tourists moved into these spaces and the flows and rhythms of the city that they interrupted and created as a result. Their analysis simply would not have been possible without the use of this more advanced quantitative technique. This kind of approach does, however, require greater technical expertise than simply watching hours of footage back and making notes about when interesting events occurred as was the case with something like Lloyd's project discussed above.

While there has been quite a lot of attention paid to the use of video within mobilities scholarship, other approaches have been used. Brandajs and Russo combined their automated analysis of video with more conventional qualitative techniques. We see a similar approach used by Birdsall and Drozdzewski (2017), although in their case the automated analysis was of ambient sound. For this project they were undertaking an autoethnographic documentation of the Silent March (*Stille Tocht*), an annual memorial parade in the Netherlands. Much of their analysis is fairly conventional, using a mix of field diaries and video footage shot while taking part in the march. In addition, however, they used Sonic Visualiser, a piece of software which helps to categorise the different kinds of soundscapes captured in an audio recording. This allowed them to quantify how the soundscapes shifted as the marchers got closer to centre of Amsterdam for the ceremony in Dam Square which forms the emotional centrepiece of the event.

Increasingly, even ethnographic research has a strong technological component. Kristian and Brady (2019), for example, discuss different techniques for online ethnography. They distinguish their approach from quantitative 'big data' analysis by talking about the need to examine qualitative 'small data' that exists online. They suggest trying different activities with participants, for example working alongside them to explore how they navigate a particular app, documenting the microgestures of swipes and taps. They also propose a 'scroll back' approach, where researchers ask participants to give them a guided tour of the traces of their life recorded on social media, positioning participants as 'co-analysts' of these datasets. Gössling and Stavrinidi (2016), conversely, explored the Facebook profiles of participants within the authors' friendship networks. They used ethnographic content analysis to code and identify themes cutting across photo albums, status updates, shares and comments for each individual. Their project specifically looked at how travel and mobility were represented within those profiles but the technique could equally be used for examining a range of issues in how individuals present their lives within social media. Compared to Kristian and Brady's careful emphasis on co-analysis with participants, there are some ethical issues to consider when treating personal friendship networks as a source of data to be exploited for research. Although Gössling and Stavrinidi secured informed consent from their participants and took care to ensure they were anonymised in publication, there is something a little discomforting about using personal Facebook 'friend' connections for this kind of online ethnography.

The PlymTour example discussed above demonstrates the possibilities of designing specific tech-led interventions to explore questions around mobilities. We should not be tempted into assuming that an intervention-based approach is an inherently superior means of creating and collecting data, but interventions do often come with a more explicitly *activist* mindset attempting to make a positive impact on the lives of participants. This can be seen in the work of Castro et al. (2016) who developed a smartphone app for use in an educational setting with at-risk teens in Quebec. One of the common issues for teens who have been excluded from conventional education settings is that their mobility is often highly constrained. The *MonCoin* ('my corner') project included an app that sent students out on weekly 'missions' exploring different civic environments both in the city and elsewhere. Students documented their missions and road trips using a shared Instagram page, giving an opportunity for creative expression. The researchers recommended the use of mobile technology not simply as a teaching tool in itself but also because of the way it helped students reframe their relationship with space and movement.

In order to segue into the next section examining practices of mapping, I am going to end this discussion by briefly reflecting on my *Rescue Geography* project. I do not want dwell too long on what is now rather an old piece of work, but the intention in creating the project was to use the mobile body as a tool for capturing an oral history of place. In what has become by far my most cited publication, we used an analysis of mapping data to demonstrate that walking in a location will elicit more explicit conversation about that place than is the case with a conventional sit-down interview (Evans and Jones 2011). Through the use of walking interviews, we created an archive of material about people's recollections of Digbeth, an area of Birmingham that was due to be extensively redeveloped at the time we were working there (Jones and Evans 2012a). The reason for discussing this project here is to reflect on some of the embodied constraints on using mobile methods. We struggled to recruit participants for this project, for a number of reasons, not the least of which was that not many people lived in Digbeth at the time and it had a somewhat threatening feeling because of the lack of people on the street. Not everyone was therefore willing to wander around talking to a researcher, even during the daytime.

In a subsequent project that I ran which used Balsall Heath as a case study, my then postdoc Saskia Warren attempted to employ the walking interview approach while working with a community of Muslim women who were learning English at a local college. Saskia subsequently wrote a nice article which was rightly critical of walking interviews as a research technique when working with people whose mobility was highly constrained for a variety of social, cultural and religious reasons (Warren 2017). Indeed, elsewhere on that project one of our participants taking part in an arts activity commented that she would be likely to face questions from her husband when she got home that evening. The reason for this was that she lived in a very close-knit community and she would have been seen walking in the neighbourhood with unknown white and African-Caribbean men from the research team; she knew that someone would subsequently call her husband to ask what she was doing. Thus, public and other spaces are a resource that is not equally accessible to all; researchers need to take this into account when designing projects including a mobile methods component.

Mapping

Many critical scholars are uncomfortable with maps and mapping technologies, not entirely without justification. The history of mapping is a history of power and control. Maps were used to carve up territory between powerful states, acting as a tool of conquest and domination. By knowing a territory, it was easier to subjugate the people who lived there, exploit their natural resources and enhance imperial power. Indeed, the very name of the UK's national mapping agency, the Ordnance Survey, highlights its military origins, mapping Scotland and Ireland in the 18th and 19th centuries to help keep the rebellious territories under English rule.

Cultural geographers are very keen on emphasising the fact that 'space' is not a set of Cartesian coordinates and that maps are merely a form of representation. Indeed, maps are often seen as problematic not just for their military origins, but because they carry an aura of scientific precision which can be used to flatten and categorise people and communities, effacing the richness and complexity of their lives. Researchers working on mapping have long acknowledged these issues, with critical scholars such as Denis Wood and J. Brian Harley framing these debates as far back as the 1980s and 1990s. Wood's most famous contribution, *The Power of Maps* (Wood and Fels 1993) explicitly countered the idea that maps were representations of objective truth, acknowledging that they always came from a subject position and were created, usually by very powerful people, with a specific purpose in mind. Harley (1989) meanwhile, drew on Foucault to talk about maps as sites of power-knowledge, arguing that maps should be considered purely as social constructions.

Acknowledging that maps come from positions of power does not automatically mean that they are therefore always *bad*. Map libraries are disappearing globally with the shift toward online mapping. Those that still exist primarily play an archival role and scholars fortunate enough to work at an institution that retains a map collection have access to what are fascinating and often very beautiful objects. The University of Birmingham has an archive of British maps from the North Africa campaign during World War II, often produced and printed in the field by military cartography units. Pre-existing maps were annotated with useful information, such as the location of defences and terrain that could be easily traversed by military vehicles (Figure 6.1). Behind these maps as technical objects are the bodies of the men who risked their lives producing them to empower others who fought and died driving back the armies of the National Socialist regime. Even overtly military maps, then, can be an agent of positive change in the world.

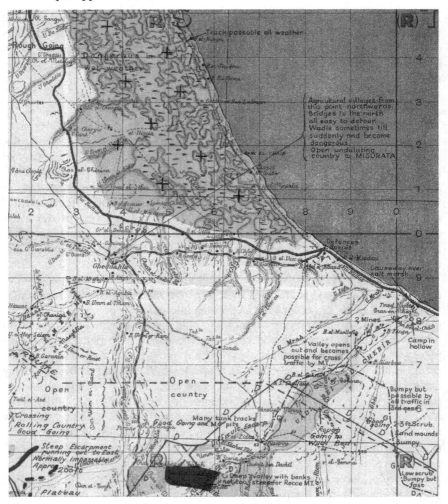

Figure 6.1 Extract of a UK military map produced for the North Africa campaign
during World War II.
Source: University of Birmingham Map Collection. Scanned by Chantal Jackson.

Nonetheless, some critical scholars seem inherently suspicious of maps as a
technology that reduces complex worlds to simple representations. Indeed, in
mobilities research while the use of video technology is widespread, many
fewer scholars in this area actively engage with mapping. Admittedly, a cynic
might suggest that learning to use mapping software is a good deal harder
than getting to grips with basic video editing. The various textbooks that have
been produced on mobilities usually include a chapter toward the end on
mapping or the use of global positioning systems (for example Ahas 2011,

Dodge 2013) but it remains an unfashionable approach among scholars in this area – though with important exceptions such as the work of Clancy Wilmott (2016a, 2016b).

Nonetheless when looking beyond mobilities studies, there is a long history of critical scholarship engaging with mapping to examine movement. Some of the most influential work in this area was produced by the Swedish geographer Torsten Hägerstrand and his pioneering 'Research Group on Human Geographic Process and Systems Analysis' established at the University of Lund in 1966. Hägerstrand's body of work tends to be reduced to his notion of space-time diagrams, a technique for plotting the movement of people and objects across a geographical region over time. To the normal XY coordinates of conventional mapping, Hägerstrand added a Z-axis which represented the passing of time (Figure 6.2). Objects and people coming together at the same place at the same time were represented by lines converging and moving upward together in what he referred to as an 'activity bundle'. As a technique,

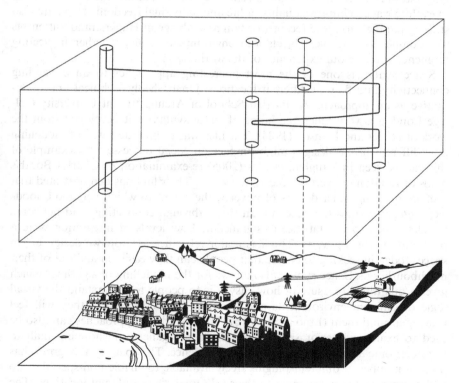

Figure 6.2 Hägerstrand's space-time cube. Although not without its shortcomings, Hägerstrand's work provides a valuable tool for examining movements through space over time.
Source: Redrawn by Chantal Jackson.

these diagrams remain a useful tool for representing movement although modern GIS software makes them much easier to create. Mei Po Kwan, for example, used space-time diagrams to illustrate the shifting constraints on the lives of Muslim women in the US in the aftermath of the 9/11 attacks. Kwan and her team undertook individual interviews with participants connecting their stories to space-time plots derived from activity diaries that they kept for the researchers. Together these created a rich understanding of what it was like to be a person who literally *embodied* the fears of Americans at that particular point in US history (Kwan and Ding 2008).

Kwan's approach of adding qualitative contextualisation overcomes one of the common critiques of the space-time diagram, that it appears to treat people as objects, seemingly lacking any sense of individual agency. Indeed, Mike Crang (2001, 194) suggested that producing these diagrams 'tends to produce a cadaverous geography. A geography of traces of actions, rather than the beat of living footfalls.' To be fair to Hägerstrand, there was much more at play in his work than simply creating a tool for representing movement. As Pred (1977) argued, Hägerstrand's concerns were humanistic, seeking to examine life-paths in order to explore ways to improve quality of life and individual freedom. There are also striking parallels to more recent work on assemblage in Hägerstrand's attempts to examine how people, objects and environments come together in fleeting moments to create our experience of the world.

Space syntax is one of the most interesting approaches to understanding connectivity and flow in different locations. Unabashedly technical and quantitative as an approach, the Bartlett School of Architecture at University College London is very much the home of space syntax as it developed from the work of Hillier and Hanson (1984). The idea was to provide a way of modelling how different urban designs might impact on social processes. An example of this can be seen in Vaughan et al.'s (2005) re-examination of Charles Booth's maps of London's poverty in the 19th century. Their findings demonstrated that even considering other drivers of poverty, the extent to which neighbourhoods were *integrated* into the wider urban tissue through connecting road networks could be seen to be statistically significant. Low levels of integration were a predictor of worse poverty. This has major relevance for how we design urban spaces today in seeking to ensure that people can flow easily in and out of their neighbourhoods. Space syntax also allows for the modelling of sightlines which gives a quantified measure of how easy it is to people to understand the visual logic of different environments and therefore how comfortable they will feel moving around them (Emo et al. 2016). These kinds of approaches can also be used to examine interior spaces such as an art gallery or shopping mall to examine some aspects of the embodied experience. The space syntax group has made a number of different plugins freely available to enable non-specialists to add a space syntax component to their GIS analysis of different locations. The rather pretty maps that result do require careful interpretation, however, and there is always a danger of overclaiming based on the apparent authority of the underlying calculations being represented in a visually compelling form.

It needs to be acknowledged that perhaps the majority of research using GIS is produced with a positivist mindset, creating simplified representations of social processes. As critical scholars have highlighted, these representations possess real power which needs careful treatment. To give an example, a few years ago I was approached by some colleagues working in social policy who were looking for someone who could make maps of data around ethnicity and poverty. I cannot reproduce these maps here because they were based on a confidential commercial dataset that Birmingham City Council – who commissioned the work – had paid to access. Experian, the credit scoring agency, will sell datasets with details about every household in a given area in the UK. This data is *geo-referenced*, tagged to the exact coordinates of people's homes. There are a number of Experian datasets available and in this case the City Council had purchased one that included detailed and highly sensitive information such as whether anyone in the household had a county court judgement against them for missing debt repayments.

Census data on ethnicity is typically aggregated to the neighbourhood scale – you can tell that there are, say, 12 people of Chinese origin living in a census tract, but you cannot tell where they live. The Experian dataset I was using had modelled ethnic origins down to the household scale, so that you could see which individual houses had people living from a particular ethnic group within them. How reliable this modelling was is open to question – my own home was apparently occupied by a Welsh woman despite my being a cis Englishman living alone. Nonetheless I was able to produce a range of maps for my colleagues presenting different ethnic groups as a scattering of household scale dots across the face of the city. Underlying these maps, I placed a fairly standard choropleth representation of the UK government's Index of Multiple Deprivation, showing the relative poverty of neighbourhoods across the city.

My colleagues had been interested in whether people from countries which had joined the European Union since 2004 (places like Poland, Bulgaria and Romania) were concentrated in particular neighbourhoods and whether local government spending on services would need to be retargeted as a result. The maps were designed to be quite a striking tool to help convince policymakers, playing on the idea that these were objective and scientific representations of the city. The analysis in fact showed no obvious patterns to where post-2004 EU migrants were settling in Birmingham – they appeared fairly well spread out and not notably concentrated in areas of the worst poverty. Making the same map for people of Pakistani origins, however, showed two things. First, there were many more households, although this was unsurprising in a city that was approximately 13% Pakistani origins at the 2011 census. Second, and more significantly, those households were overwhelmingly concentrated in the areas of worst poverty in the city.

Clearly, there are many economic, social and cultural reasons why even second and third generation Pakistani migrants might still be living in poorer neighbourhoods of Birmingham (Gale 2013). I use this particular map with my students as an example to illustrate how these powerful representations have the potential to be misused. If presented to the City Council, such a map might make an argument for targeting greater investment in transport, education and training in these neighbourhoods, developing initiatives that take sensitive account of cultural practices among those communities. Presented to a group concerned about the impacts of immigration or simply to a racist political party, these maps could be twisted to tell a story of how Muslim groups 'create' poverty and thus contribute nothing to the UK economy.

The point of reflecting on this project is to demonstrate how maps can be used to impose a particular narrative on a community or neighbourhood, framed by something that looks like an objective, scientific truth. Critical cartographers have been working for decades to find ways of subverting the power to impose narratives from above, giving communities more control over the stories that are told about them through mapping. From the mid-1990s scholars working in Public Participation GIS began to problematise the uneven relationship between the community at large and those professionals with access to high-end GIS software capable of sophisticated analysis and display of data (Obermeyer 1998). By the early noughties there were a number of complementary initiatives working in this area variously labelled Participatory GIS, counter-mapping, GIS and Society, GIS/2 and critical GIS (for an explanation of these different terms see Sieber 2006).

The early years of this movement saw researchers relying on their own specialist GIS systems to actually do the mapping for the communities they were working with. Indeed, Carver (2001) was lamenting the lack of online systems for implementing Participatory GIS, just as Crampton (2001) was predicting that hypermedia and distributed mapping via the web would be the future for more participatory and critical mapping. Crampton was not wrong and the speed at which the tech sector subsequently upended conventional mapping practices is quite shocking to look back on. Google Maps was launched in February 2005 and, unlike previous efforts at online mapping such as streetmap.com, was much more than simply a means of accessing conventional maps through a website. As Zook and Graham (2007) noted at the time, Google Maps and similar products from Apple and Microsoft, created a hybrid space where data about physical locations could be digitally searched. This also offered the potential to add layers of user-generated content onto a base map to create new, multimedia, representations of space.

Thus when Cope and Elwood (2009) produced their edited collection *Qualitative GIS*, they were reflecting on what was by that point a well-established field which was being turbocharged by the activities of the tech sector. We started to see examples of programmers, rather than geographers, taking these online map services and creating mash-ups (a very mid-noughties term!) by overlaying data from different sources. Following Hurricane Katrina, the site scipionus.com was set up by two software engineers based in Austin. Although it only existed briefly, it allowed survivors to post markers onto a Google Map. Simple text strings could be tagged to a geographic location to tell a crowdsourced story about the banality of tragedy, with people doing things like marking their house to tell others that they got away safely or trying to track down the owners of pets found abandoned at a property (Miller 2006).

Another early map mash-up had a direct influence on policy. Chicago-Crime.org was set up by Adrian Holovaty to bring together the city's crime data which was made publicly available by the police. Holovaty created a user-friendly interface to map where the city's crime was concentrated. The UK's Conservative Party, then in Opposition, explicitly cited the potential of Google Maps to empower residents by providing them with the same kinds of spatial information available to local politicians and the police (The Conservative Party 2008). Returning to power in 2010, the Conservatives made good on a promise to make much more government data publicly available, including police data on reported crime.

As discussed in Chapter 2, simply making data open for anyone to download does not in itself redress power imbalances – indeed it can reinforce them. By making information accessible to privileged groups who have the resources to analyse and interpret these datasets, the powerful are able to lobby for policy changes that protect their interests. Indeed, this issue of needing third-party facilitators with mapping expertise to work with communities was identified as a key concern during early debates about participatory approaches to GIS (Elwood 2006). This is an issue I am fairly familiar with as I am periodically approached to map various datasets for different organisations that lack the skill to do so. Recently, I have worked with the University of Birmingham's Student Services department to examine the level of crime in Selly Oak, a studentified neighbourhood adjacent to the main campus.

Crime is a big concern for students, particularly those attending university in larger towns and cities. Part of the reason I was asked to examine and map crime data for Selly Oak is because of a mismatch between student perceptions of high crime rates and a discourse from the West Midlands Police that this is a low crime and therefore non-priority area. UK crime data from 2016 onwards is publicly available for download, giving information on the number and types of crime committed and whether there had been any outcome such as an arrest. Each reported crime is also linked to a geolocation, aggregated to the level of an individual street (for example 'On or near Raddlebarn Road'). This goes some way to protecting individual privacy – people may not wish their neighbours to know that they were sexually assaulted in their home for instance – though can make the maps look a little unusual, with multiple crimes seemingly committed in an arbitrary location in the middle of a street (Figure 6.3).

The Student Services department hoped that by providing a more detailed breakdown of where crimes were taking place in Selly Oak, the times of year where they peaked and the types of criminal activity involved, the university would be able to reassure students that they were not really in any danger. Indeed, by mapping the data at street level it is relatively simple to exclude

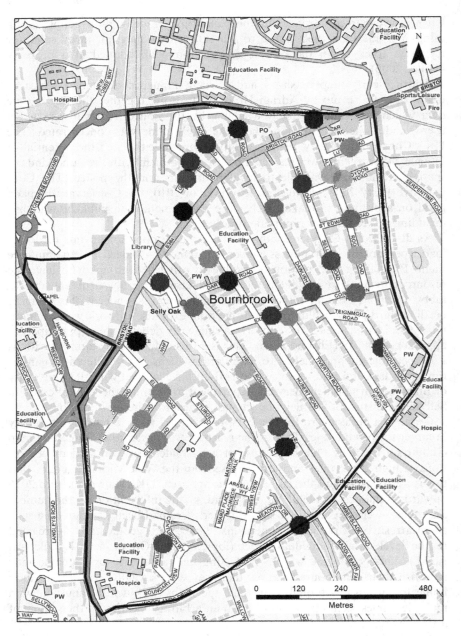

Figure 6.3 Crime levels in the Selly Oak neighbourhood of Birmingham. Based on open source police data from 2016–19, the map was produced in response to student perceptions of the district neighbouring the university campus being particularly high in recorded crime.

Source: Author's analysis of data from data.police.uk. Contains OS data © Crown copyright and database right (2019).

crimes that are not a *specific* threat to the student body, such as the large amount of shoplifting that takes place at a nearby retail park and the number of assaults that occur in the accident and emergency department of the local hospital. This filtering shows antisocial behaviour and assaults unsurprisingly clustered on a street of bars and restaurants in Selly Oak. Burglaries, meanwhile, reach their peak in October and June when students are busily moving in and out of their rented houses and are less alert for strange activity in the neighbourhood. Purely looking at reported crime, then, Selly Oak is not a particularly dangerous place.

The student perception of Selly Oak is, however, very different. A stabbing took place in the neighbourhood after an attempted car-jacking in 2018, with the 15-year-old perpetrator caught and subsequently jailed for 18 years (Woodley 2019). For some residents, the incident was taken as being symbolic of the daily dangers that students face in Selly Oak, with stories of young women being followed home, threats and abuse being common currency. Some of this can be dismissed as being a product of an oversensitive, privileged and overwhelmingly white student body which grew up in leafy suburbs. Place a large number of such people in close proximity with access to social media and rumours unsurprisingly spread like wildfire. None of this is, however, any comfort to the young person who has had abuse shouted at them while heading home and who has known someone be a victim of physical or sexual assault in their neighbourhood.

There is potentially a positive story here about using mapping technologies both to provide some reassurance to students and to press the police and local authority to divert resource to tackle specific problems that the student body faces. This might be, for example, funding CCTV systems in streets that can be identified as particular hotspots. The police themselves are unlikely to have the resources to examine a neighbourhood with this level of granularity, particularly when the aggregated data demonstrate that this simply is not a high crime area or a priority in an age of overstretched budgets.

The less positive interpretation of this, however, is that an already highly privileged group (white, middle class, with high levels of educational capital) is being given the tools to lobby for greater protection from criminal activity. Although that crime is doubtless distressing at a personal level, events in Selly Oak are frankly not that serious compared to other neighbourhoods in the city that have been ravaged by decades of poverty, violence, drugs and gang activity. Is my activity as a researcher here simply helping what is a wealthy and powerful institution to lobby for greater resources for its students from a cash-strapped local authority and police service? In doing so am I complicit in a neoliberal agenda that protects the interests of the wealthy by using expert knowledge to draw on resources that could otherwise be spent on society's poorest and most vulnerable?

Location-based services

In 1978 the first launch took place to start setting up the US military's global positioning system (GPS) which placed a web of satellites in geosynchronous orbits surround the globe. Based on some fiercely complicated maths, by calculating the time that signals are received from these satellites a device on the ground can work out its position through triangulation. Until 2000 the original US system introduced deliberate errors into the signals accessible to civilians but it has subsequently become the backbone of a huge number of industries and consumer applications. Rival systems such as Russia's GLOSNASS, China's BeiDou and the EU's Galileo array remove the dependence on a US system that could be switched off at any time in the event of military tensions.

While the iPhone was not the first smartphone, it doubtless kicked off a revolution in making mobile computing ubiquitous. A GPS receiver was not included until the launch of the iPhone 3G in 2008, before which users were only given a rough approximation of their location generated by triangulating the signal from different cellphone towers. Very quickly, however, GPS chips were included in all smartphones allowing location-based services to take off.

In 2008–9, just at the tipping point where apps on mobile devices were becoming established, I ran a couple of location-based projects using the now obsolete Windows Mobile platform. In one of these I built a simple logging tool that allowed walkers to press one of four buttons on a handheld device depending on how happy or unhappy they felt at any moment while walking through a neighbourhood. In the other, I commissioned one of my students build an app which allowed users to take geotagged photographs and give them a ranking, creating a colour-coded map which could be viewed in Google Earth (Jones et al. 2011). Both showed the way for enabling participants to help researchers access local knowledge about different spaces in a neighbourhood. Although cutting edge in 2008–9, both projects seemed comically crude within a very short space of time as the sophistication of mobile devices and their associated software ramped up extremely quickly. As Dodge (2013, 527–528) put it a few years later:

> Many people today are carrying around mobile devices with sophisticated mapping capabilities that were the preserve of the best-equipped elite military forces only a decade or so ago. This is remarkable because most people regard it as unremarkable!

As a result, arguably many people today are *doing* qualitative GIS without even being aware of it. The major social media platforms (Weibo, Snap, Instagram, Twitter, Facebook and others) all give users the option of posting their location, creating a treasure trove of geotagged materials for these companies. Goodchild (2007) predicted that these new technologies would turn citizens into *sensors*, allowing for the crowdsourcing of a host of interesting geographical information. Reducing the body to a mere sensor to create data is of course problematic as it effaces the identity and agency of the individual. Still more concerning is the fact that these data are being harvested by massive global corporations who exploit them for commercial gain.

Some of these datasets are made available to researchers in limited forms or for a significant fee, but for the most part they are treated as a commercially confidential asset. Even the users who generate these data effectively surrender their rights to it. Rather than dwelling on this somewhat depressing aspect of geolocation, the remainder of this chapter explores some of the interesting research applications that have been enabled by the fact that what was once military-grade geolocation has become a ubiquitous and invisible part of our lives.

Location-based services have created a whole range of new ways to interact with the world. A game like *Pokémon Go* only exists because of the way a smartphone can combine a real time data connection with a GPS signal that allows a game world to be laid on top of our physical environments. Other services rely on GPS in less visible ways. Some insurance companies now require young drivers to fit GPS-based black boxes. If these detect that the owners are driving too quickly (or even, in some cases, after a certain point in the evening) the insurance can be invalidated or higher premiums charged (Derikx et al. 2016). Meanwhile, most medium and larger-sized consumer drones have GPS capability built in. This allows these drones to travel along pre-determined flight paths, hold station in particular locations and even track moving objects in a semi-automated manner, reducing the need for embodied skill from the operator. Geographically sensitive locations such as airports can be geo-fenced within the aircraft's software to prevent users accidentally or maliciously overflying these spaces. It is, however, a relatively simple process to by-pass these restrictions and drones have become a key vector for blurring the boundaries between inside and outside in prison spaces. Drugs and mobile phones can be flown into prisons, with drones combining GPS and onboard cameras to precisely locate and hover outside a specific window, allowing prisoners to simply reach out to grab suspended packages of contraband (Rubens 2018). Via a colleague I was briefly involved in a discussion with UK's Home Office about the problem of flying drugs into prisons. This included mapping the potential flight range and exploring the problems of trying to remotely shut them down. Drones broadcast on the same frequencies as domestic WiFi interfering with which could be highly disruptive in built-up areas as could attempting to jam their GPS signals given the number of devices that now rely upon it.

Location data can thus be used for a variety of purposes, many of which are very practical but there are also aesthetic applications. One can see this approach in the now fairly common practice of GPS writing, where people draw out words and patterns via the track of their movements around a city. My 'RIDE' map discussed in Chapter 5 is an example of this kind of practice. Such activity allows us to see urban spaces differently and the potential they have to have value beyond facilitating economic exchange. A number of artists have undertaken interesting projects that play with location and mobile devices. The paintersflat group, for example, created a hiking trail linking two World War II bunkers through the Silver Island mountain range in Utah. They took a digital terrain model of the area and used the kind of generative algorithm used in video game design to plot a route between the two sites.

Hikers could download this as a GPS tracklog and 'walk like a bot' across this hostile and beautiful desert landscape (paintersflat 2007). Such work asks questions about how we access different environments and the degree of agency we have in our everyday interactions with them. It also demonstrated how accurately the modelling software reproduced actual human mobilities, with parts of the computer-generated trail coinciding with existing roads and paths, reflecting the most efficient route across the hills.

Wilken (2010) has reviewed a number of artistic interventions using different kinds of locative media including the Jabberwocky app which allowed users to see when they were within range of a Bluetooth enabled device that the app had previously tracked, allowing users to uncover familiar strangers in their everyday spaces. This idea of being able to connect with nearby strangers is, of course, at the heart of many dating apps. The 'swipe logic' of Tinder gives users access to a carefully curated window into the lives of people in their locality (David and Cambre 2016). Individuals are reduced to a few pictures and a small number of words with the online platform offering a way to connect to the physical self. Gay dating app Grindr is even more overt in the way it uses location data, giving users the opportunity to see how many potential romantic partners there are in the vicinity and how far away they are. Grindr uses a virtual interface to layer 'gay space' on top of material spaces to create opportunities to easily meet other gay people without needing to go to a location catering primarily to a gay clientele (Blackwell et al. 2014). As such these location-based services are driving significant changes to the way that physical spaces appear and are used.

Given the sensitive information held within the servers of dating apps, there are, however, significant questions over security. Multiple dating apps have reported data breaches, for example the 3fun website having a vulnerability revealed in 2019 that exposed not just personal details but GPS location for all 1.5m users, which could be accessed in real time by malicious hackers. Given that the service is dedicated to helping people find others interested in group sex, this is particularly problematic in terms of the potential for blackmail. Indeed, there have been concerns raised because Grindr is now owned by a Chinese company, which opens the service to national security concerns for non-Chinese users (Porter 2019). Location data thus adds an explicitly embodied dimension to the privacy concerns discussed in Chapter 2 because it becomes possible to physically find the individual in real time, not simply look up information about them.

In reflecting on research methods based on location-aware technology, my *Rescue Geography* project discussed above was entirely reliant on GPS tracking to link the stories told in walking interviews to the locations where people were talking about different issues (Jones and Evans 2012b). Since that project, Wilmott (2016a) has undertaken a fascinating and nuanced critical analysis of just how the GPS data recorded during mobile interviews can be deconstructed to examine the process of producing and exploring these datasets. She reflects on some of the behind the scenes processes that go into producing the maps that we use in our analysis and publications, showing how columns of numbers in a spreadsheet are transmuted into a trace of a lived experience. Indeed, she argues that the rapid growth of mobile devices with mapping capabilities has changed the nature of debates within cartography. GPS tracking has brought an explicitly temporal component to mapping just as the spatial turn within the humanities has brought the consideration of space to the exploration of video and other visual texts.

As GPS-enabled devices have become more common, so an interest in their research potential has started to spread beyond the discipline of geography. One of my former PhD students, Tom Disney, was employed on a project looking at the lived experiences of social workers. This led to my being brought in as a consultant for a team of social work scholars who were interested to see whether there was any research potential in looking at the travel patterns of practitioners. Social work is a highly mobile practice, particularly when working with children, where practitioners have to drive between visits to homes and schools, arrange supervised meetings at neutral locations and undertake a host of other activities as well as working around their own personal travel demands.

Tom asked a group of practitioners to carry a GPS logging device for several days, then had them talk through the resulting maps of their daily journeys. I established an analysis protocol for him and helped out on processing some of the data and in return he kindly named me as a co-author on the subsequent paper (Disney et al. 2019). It was an unusually depressing project to collaborate on, with the data giving a window into the lives of practitioners working with some of the most challenged and challenging families in the UK. For all that Crang argues that the traces of movements in space-time diagrams are 'cadaverous', in fact one can read quite a lot of detail simply by examining where people go and how long they spend there. Indeed, using Google Street View, it is even possible to get a sense of the actual sites being visited. Even without the specific contextual information from the interviews, it is quite easy to infer what is being represented in these movement logs for a researcher with just a basic understanding of what the job of a social worker entails: the repeat visits to houses in some of the poorest neighbourhoods; the practice of parking some distance away so that the vehicle is not attacked; the meetings in schools followed by journeys back and forth to parental homes; the visits to court; the extended round trips to children who have been placed a considerable distance away from the social worker's 'patch'. Even the fact that many of the practitioners were routinely leaving home very early and returning quite late gives insights into just what a challenging sector they are working in.

As a side note, this project also highlighted some of the simple technical challenges of using location data in research. Given that this data revealed the home addresses of both practitioners and service users, it had to be treated as highly confidential and sensitive, with great care taken both over how the data were transferred between researchers and to ensure that published maps were sufficiently abstract to prevent locations being identified. Another crucial issue is that GPS is very power hungry and, if left permanently on, can quickly drain the battery of a mobile phone. For the social worker project, Tom used a standalone logging device with a large battery but was reliant on his participants to charge it overnight which they often forgot to do, thus losing any data the following day. GPS is also no use for tracking people inside buildings and so smaller scale movements of participants around an office, say, cannot be recorded this way. Indeed, in areas with tall buildings, GPS signals can bounce around giving inaccurate or sometimes incomplete logs of movement. These limitations are very important to consider when designing a project using this technology. Indeed, participants will not thank you if the app you ask them to install on their phones for you drains their battery in a matter of hours!

Nonetheless, it is considerably easier to ask people to somewhat passively gather spatial data for you than it is to empower them to gather and analyse their own material to create maps themselves. Accepting that limitation, however, it is still possible to do projects that seek to act in the interests of communities while using members of those communities in a 'sensor' mode rather than as active co-constructors of the research. A nice example of this is the work of Yip et al. (2016) examining residential segregation in Hong Kong. They built an app that only activated the GPS sensor of participants' phones once every five minutes, thus keeping the impact on battery life to a minimum while creating a daily movement log. For their 71 participants they were able to record 1,073 individual journeys, showing the amount of time spent in different kinds of neighbourhoods. A key finding was that people from poorer neighbourhoods spent time on social activities in other similarly poor neighbourhoods. Thus, despite the compact nature of Hong Kong, the data showed relatively few locations where richer and poorer communities would come into social contact.

The findings from a project like this could be used to feed into a policy discussion about the nature of segregation and socialisation. Whether policymakers would be keen to do more than pay lip service to breaking down segregation at the wider society scale is, however, quite a different matter. Nonetheless, there are interesting opportunities to feed into policy debates using location-based data. An example of this can be seen in a project I ran that responded to the Localism Act, 2011, which created powers for communities in England to take greater control over the planning of the neighbourhood in which they live. One of the issues with this legislation was that in practice the powers were very limited and favoured communities that already had or could afford to buy-in expertise to deal with the technical aspects of compliance with national, regional and local planning frameworks.

We created and piloted a tool that could be used to help communities overcome at least some of the barriers to gathering the necessary information to start putting a neighbourhood plan together (Jones et al. 2015). Developing from my earlier project building a geotagged photo logging tool, we created an app for Android phones that allowed participants to take geotagged photographs, notes and voice memos which would be seamlessly uploaded to a shared community map. Because audio clips were used, individual participants could not be anonymised meaning that in order to generate informed consent there needed to be careful explanation of the fact that the data would be open for anyone to examine. The 50 people who took part in the pilot generated over 1,000 photographs and 626 audio clips for the two case study neighbourhoods that we were working in. Everyone involved had access to the whole dataset and it could be freely downloaded from the project website to be viewed in Google Earth.

Gathering data about a neighbourhood is very much the easy part of putting together a local plan and in this small pilot we did not even begin to move from crowdsourcing the data *collection* toward crowdsourcing the data *analysis*. Nonetheless there was some interest from local policymakers in the potential of such a tool for capturing residents' ideas for how to re-plan a neighbourhood, identifying specific problems that needed to be addressed and highlighting areas that might be the priority for action. A much more sophisticated app along similar lines has since been developed by my colleague Sophie Hadfield-Hill as part of a suite of techniques within her work on child-friendly urban designs in the developing world (Hadfield-Hill and Zara 2018). Her work has really demonstrated the power of this kind of approach and, indeed, the project findings have generated significant interest at both national and international policy scales.

Conclusion

Using a GIS to undertake even basic mapping tasks can quite tricky. As a piece of software, it seems to generate strong emotional reactions from students in classes, creating irritation and outright fury, frustration and even tears. It can also be quite magical, as a dataset that seems entirely abstract when viewed in a spreadsheet suddenly gains meaning when placed in a spatial context. There is, however, much more to mapping as an approach than working on a user-unfriendly desktop GIS. Indeed, many of us create and use spatial data every day without necessarily even realising it.

I have placed mobilities and mapping together in this chapter because I think they have a natural affinity even if many mobilities scholars seem a little reluctant to engage with mapping technology. Nonetheless, these two areas of research should be considered as complementary rather than interdependent and both fields offer tremendously interesting possibilities for research in their own right. In terms of the discussion in this chapter, these can be summarised as:

1   projects examining the power relations embedded in the different technologies that have become entangled with mobilities practices;
2   projects that explore new approaches to investigating mobilities beyond ethnographic and video-based techniques;

3   projects that use maps in ways that undermine narratives of place imposed from above;
4   projects that use mobile technologies in a more interventionist manner to help change how people relate to the spaces around them.

As someone who has worked on the fringes of mobilities scholarship for many years I find much to be excited about in the ways that some scholars are pushing the research agenda in this area. For all that I might sometimes be a little cynical about the occasional tendency for some scholars to produce an overtheorised account of a relatively banal travel experience (and I will hold a hand up to doing this myself, e.g. Jones 2005) there is some really great work being done that stretches across disciplinary boundaries. From a technology point of view, some genuinely impressive work has been made possible because of the precipitous decline in the cost and size of broadcast quality video cameras.

Newly available technologies bring problems as well as opportunities to researchers, however. Because video storage is now incredibly cheap, researchers can end up with hundreds of hours of footage that require careful thinking about how to analyse. Automated approaches using computer vision techniques will be useful for some kinds of project and allow elements of quantification to be introduced though many critical scholars would have to seek out an expert collaborator, at least to get started on an analysis of this kind. Such approaches are unlikely to capture the kind of micro-interactions that many mobilities scholars are interested in where a very time consuming, manual analysis may be the only option. As always, forward planning is essential.

Researchers who learn even quite basic GIS can find themselves in high demand from colleagues and external organisations who want a mapping component to their data analysis. One of the reasons why this skill can be in high demand is the fact that a map brings a compelling aura of scientific authority which can add weight to project findings. It must be frankly confessed that a great deal of work involving mapping is unashamedly positivist. Nonetheless, critical scholars have spent many years finding ways to subvert the apparent authority of maps, giving communities and others the power to write their own stories into these socially constructed representations of space. Even methods which on the surface appear quite deterministic such as space syntax, will often have a great deal of critical thinking underneath them as well as a socially aware agenda.

Around 2010 it became something of cliché in grant applications that people said that they were going to build an app. Funders quickly became more than a little cynical about whether these apps would actually be of any use. The rise of location-based services on GPS-enabled devices has, however, undoubtedly changed many aspects of the world around us as well as approaches to research. Things like dating apps have, very simply, altered how people are able to *see* the physical world around them, with new information being layered into the world they are occupying. As researchers, being able to locate a body in geographic space allows us to plot movement with a precision that was hitherto impossible. Activity diaries remain a fantastic research method but getting people with mobile lives to talk through a GPS track of their movements, being able to see how much time is spent in different locations, even being able to explore something of those locations via Google Street View is quite transformative.

Thus, it would be a huge mistake for researchers to somewhat snobbishly dismiss the potential for using mapping techniques as being simply positivist, power-laden and *bad*. Of course, they can be used this way, but they can also be so much more than this. Mobility is a hugely significant part of contemporary society and, whether we like it or not, location-based services underpin many aspects of our lives. As a result, there is much work to be done investigating the methodological as well as socio-cultural implications of these relatively recent transformations to our society.

7   Conclusion

This book covers material that is close to my heart. While it has been tremendously satisfying to write, I am open to criticism that in focusing on my personal interests I have missed some significant topics in the field of technology and methods. I have not written a great deal here, for example, on big data, or artificial intelligence and there would be interesting things to say about military hardware, drones, self-driving cars, space exploration and a host of other topics all of which would warrant a volume on their own.

The point of this book was not, however, to attempt the impossible task of giving a comprehensive overview of the rapidly evolving tech landscape; I wanted instead to illustrate an *approach* to research. In the introductory chapter I described this approach as 'playful methods', but this is just an arbitrary label for embracing enthusiastic experimentation. The focus of this book has been to promote a willingness to explore the kinds of tech-led research methods that are becoming more readily available to critical scholars. Hopefully readers will develop the confidence to try some of these things for themselves if they have not already done so. Playing around with different approaches allows us to think about how these techniques might be adapted for projects within our own fields of interest, allowing us to answer new kinds of research questions.

If there is a common thread in the methods that I have discussed, it is that they allow us to address issues that are increasingly relevant for a large number of critical scholars based in different disciplines. Self-driving cars may raise a lot of interesting questions for the future, but they are not yet part of our day-to-day experience of the world in a way that could shape a new set of methods, unlike, say, GPS-enabled smartphones. In the techniques that I have highlighted here, the last two decades have seen what were once impossibly sophisticated technologies being democratised, made everyday and unremarkable.

This democratisation has changed the way that we do research. When I started undertaking archival work during the late 1990s, most scholars were still making handwritten notes in pencil and only occasionally taking expensively produced photographs of key sources. Today, laptops and smartphone photographs are the norm in archive reading rooms, allowing material to be gathered in an ever more efficient way. Indeed, in some cases archives are themselves being digitised and made accessible online so that one no longer even needs to travel to a far-flung location to access them. I do not believe that many of us hanker for the good old days when this kind of research was considerably more difficult.

There is an important caveat here. Even though many of the techniques I have talked about here are becoming more accessible, most still require a fair degree of technical expertise, often with a steep learning curve. Some scholars may not be attracted to the prospect of fiddling with an unfamiliar device or user interface, particularly when it is not immediately obvious how to interpret the data that emerge. Indeed, I argue that one of the reasons why digital video has become such a popular tool within research is because it is fairly simple to use (point camera, press record) and interpret (skim through footage, note what happened). The same cannot be said of even fairly basic mapping systems, let alone using advanced physiological measuring devices, programming within a games engine or many other methods that I have discussed within this book. As a result, *play* also requires *work* – trying things out while learning the underlying principles.

I am a great believer in learning by doing both in my research and teaching practice. This links to another pillar of the playful approach which is that the pay-off for trying out a new technique may not be in generating immediately useful research data. Indeed, the learning process itself may be more valuable because of the way it allows scholars to have more informed conversations with experts in different disciplines. Knowing what is possible widens the scope of projects that we can develop, even if this means subsequently finding an interdisciplinary collaborator who has the necessary expertise to use a technique to its fullest. Working with eye tracking, for example, is likely to be beyond most critical scholars, but knowing about these techniques and seeing what they can do could become the start of co-constructing a project with an expert in that field. My experience of interdisciplinary scholarship has always been that you learn things very quickly because what seems novel to one discipline is familiar and routine in another, which can lead to very interesting exchanges if both parties have an open mind.

As critical scholars we can identify two key problems with some of the methods that I discuss here. The first is that they can be reductive, shrinking our complex embodied lifeworlds into overly simplified digital measures. The second is that these technologies have been built by people with particular subject positions and assumptions, embedding significant problems of power and privilege. These two problems are interlinked but there are opportunities to both work around and subvert them. Many technologies emerge from a masculinist mindset that tends towards quantification, with the reassuring aura of authority that comes from seeming precision and calculability. One way to counter this is to examine the limits of what these representations are able to show through speaking back to dominant discourses using the same techniques that produced them. An example of this might be to help communities to make their own maps to counter a narrative that labels their neighbourhood as a crime hotspot. Alternatively, we can fold these technologies into mixed methods approaches, using qualitative measures to provide context to technology-derived, quantitative data. We saw this in the example of interviewing participants alongside taking measures of their physiological responses to different urban spaces.

Unequal power relations are inherent to all research methods and the job of critical scholars is to find ways of minimising the impacts of these and at the same time ensuring that participants are not put at risk. Even when considering something as basic as a qualitative interview, we tell our students that it is imperative to protect participants by thinking about where the interview takes place, the language used in the questions, the kinds of things that are and are not appropriate to ask. The need to ensure that safeguards are in place to counteract uneven power relations is no different with any of the technology-led approaches described in this volume. We have to acknowledge, however, that male and white privilege are particularly present within the tech sector which requires us to think *even more* carefully about the kinds of protections we put in place when we use different technologies within our research.

As is apparent throughout this book, the broader tech sector really does have a problem with women and ethnic minorities. The better-paid engineering and leadership roles in tech companies are dominated by white and Asian men, locking out women as well as black and Latino populations. Combined with the discourse of unfettered capitalism that runs throughout Silicon Valley this can lead to some deeply morally questionable decisions being made. Facial recognition and tracking technology, the blinkered response to the rise of body issue problems among teenagers on social media, collaboration with Immigration and Customs Enforcement on technologies to help expel immigrants and refugees and a host of other issues have rightly been flagged as symbolic of the sector's casual disregard for social consequences. In fairness, many working within the tech sector such as 'Googlers for Human Rights' have protested at some of the more dubious sales contracts being pursued by their employers and in some cases companies have backtracked (Helmore 2019).

A key tool that scholars use to protect participants is confidentiality – anonymising our data so that it cannot be tracked back to an individual. One of the most shocking trends in the last two decades has been the quiet erosion of privacy as a fundamental principle of a healthy society. I explored some of these issues in Chapter 2 although privacy issues cut across the whole book. The tech sector apparently sees nothing wrong with hoovering up huge amounts of information directly linked to named individuals. Indeed, we see this in pithy soundbites claiming that data is the new oil – a commodity on which the global economy depends that is being ruthlessly exploited for profit (*The Economist* 2017). This discussion of data as a resource casually ignores the fact that these data represent the lives of people, harvested and controlled by companies who give no meaningful protections to those individuals.

These actions reduce the individual body to its value as an advertising sale: a person's likelihood to buy a particular product or shift their vote to a rival party. We are being sold convenience at the cost of privacy. This has reached an extreme degree with the rise of digital assistants where we are literally allowing global corporations to listen into the activities within our homes. The more we consent to this, the more that such systems start to become compulsory rather than optional. The erosion of privacy has, however, generated a treasure trove of data and tools for data collection which, as I discussed in relation to facial recognition, could absolutely be put to a socially beneficial uses by a critical scholar. This means that we end up with ethical dilemmas where we need to consider trade-offs between risk to participants and the potential research gains to wider society.

Related to privacy are questions of transparency. A discourse has arisen that transparency is inherently good because it empowers people to make informed decisions. But this assumes that people are equally able to access, analyse and employ these data which is, of course, nonsense. A great deal of knowledge and skill is required to do anything meaningful with these datasets and it tends to be already privileged groups who gain the most from data transparency as it empowers them to lobby for their interests over society's more vulnerable groups. Nonetheless, very large government datasets have now been made public from crime statistics through to house sales. This opens up interesting possibilities for critical scholars to re-analyse these data in ways that can help less privileged communities to mobilise around particular issues.

Critical scholars working *with* community groups to examine newly available sources of data is one way to begin a process of empowerment, in direct contrast to the tech sector's default approach of treating people as resources to be harvested. The lack of protection to the individual data producer become even more problematic when considering the kinds of body-monitoring technologies that we discussed in Chapter 3. The discourse of the quantified self promises users *control* while passing intimate details to companies to use as they see fit. Self-measurement is rooted in the neoliberal mindset that runs rife in Silicon Valley, pushing responsibility onto the individual rather than seeking to tackle wider structural factors driving inequality and poor health. Biohacking sits at the extreme end of this pursuit of self-regulation and it challenges us methodologically by pushing into territory that undermines the principles of ethical research and participant safety that are at the heart of the academic sector.

At the same time, however, innovations driven by the tech sector have allowed new kinds of measuring devices to be manufactured. These take the physiological measures that people working in psychology and marketing have been using for many years and have made them cheaper, more conveniently packaged and with considerably more user-friendly interfaces. This makes the use of physiological measures more appealing to scholars working beyond these disciplines. Of course, one of the appealing qualities of these increasingly lightweight and discreet devices is that they tempt us toward using them outside the controlled confines of a lab setting. Doing projects in the messy, complex real world does, however, make it much harder to interpret physiological data because there are so many environmental factors to take into account. This is where mixed methods approaches including a qualitative component can be so helpful because they can give a sense of what is driving a person's physiological responses.

We can also borrow some of the interpretative frameworks from the disciplines that have pioneered the use of these physiological measures. While some critical scholars may wince at the idea that emotion can be calculated, certainly these datasets allow us to *infer* things about a participant's emotional state. Again, however, there is quite a big jump between getting a participant to strap on an E4 wristband or similar device and being able to meaningfully analyse and interpret the data that result. Some techniques, like the clinical end of eye tracking, really do require collaborating with an expert or spending a lot of time learning how to use these devices properly.

Eye tracking has a number of applications, including within the videogames sector. As we examined in Chapter 4, scholars in a variety of disciplines are increasingly taking videogames seriously. Methodologically, games are interesting both as texts to analyse and as tools for working with participants. From small developers making simple mobile games, to teams of thousands working on triple-A titles, games and the gaming sector are highly diverse. Traditionally, the sector has had significant problems with gender and ethnicity both in characterisation and the themes explored within games. These issues have not gone away, though some of the more obnoxious manifestations are slowly being eroded with, for example, storylines that do not simply treat female characters as props or eye candy. Nonetheless, there are some truly despicable sexist and racist discourses within parts of the gaming community that can, on occasion, make this a toxic area in which to do research.

Gaming practice is highly embodied and multisensory, which offers interesting opportunities to explore how questions around inclusion and exclusion play out in virtual spaces. The worlds within games themselves have become increasingly sophisticated and give a compelling sense of being *immersed* in a world; this quality can be used to test a range of scenarios by simply asking participants to play a carefully selected part of a game. With virtual reality, the sense of immersion is even more compelling and, again, this offers researchers the potential to explore embodied responses to virtual scenarios. For those scholars willing to spend some time learning the basics of using a games engine, it becomes possible to design specific experiences for participants that respond directly to a research question, rather than fitting around the confines of an existing game environment.

Moving more into the design end of gaming research fits with the themes of creative practice that were explored in Chapter 5. I have made an argument that one should consider creative practice to be broader than conventional arts-based methods, particularly in moving away from the idea that any kind of conventional aesthetic quality is necessary or desirable within outputs. There is, of course, productive potential in working with artists, both while working as a junior partner helping them realise their aesthetic vision or by asking an artist to bring their different perspectives to projects that the researcher leads. Artists often work on the cutting edge of what technologies can do and can point the way toward methods that researchers may subsequently wish to employ. ORLAN's pioneering 1990s work using video streaming and remote audience feedback, for example, is a lesson that those of us still jetting around the world to work with participants and to attend conferences could well reflect upon, particularly now that these technologies are so much more readily available.

Creative practice comes with ethical questions for researchers. The sex industry in its myriad forms is doubtless creative but is more realistically an interesting *topic* for research rather than a source of *methods* for most scholars. Similarly, while we can undertake cute, subversive interventions such as AR graffiti, one has to reflect on whether the energies of critical researchers would be better spent in more direct challenges to existing power structures. Some of the work undertaken within hackathons is very interesting in this regard, despite the uncomfortably masculinist origins of hacker culture.

Chapter 6 brought together two complementary areas of research, examining our everyday mobilities and shifting practices of mapping. Mobilities researchers have developed some wonderful techniques, particularly in the application of (auto)ethnography and video. High resolution video and very cheap storage makes it very easy to collect a huge amount of material and this raises crucial points about forward planning in terms of the analysis strategy employed by the project team. Newly emerging techniques around automated video analysis have interesting potential for certain kinds of projects but they do require more technical expertise than simply skimming through hours of footage.

It is also important to remember that mobility is not a resource that can be equally accessed by all people. Using a range of technologies to explore everyday interactions with a neighbourhood may be a moot issue if participants have severe constraints on when they can be in particular spaces and who they can be seen in public with.

Mapping technologies are based on power and dominance, from their military and colonialist origins to the ways that simplified narratives can be imposed on a group of people purely because they live somewhere that has been labelled on a map as being in some way problematic. Some critical scholars are uncomfortable with mapping, with many geographers protesting that there is more to space than representations of Cartesian coordinates. Of course, this is true, but there is a very long history of scholars working in cartography and mapping who seek to subvert the power of maps. Critical and qualitative GIS find ways to present different narratives about people and places, empowering communities to resist the totalising narrative of official mapping. Nonetheless, those of us who can use GIS software must be mindful of how we collaborate with different organisations and the uses to which the maps we produce can be put. Maps can create powerful narratives and we need to be careful that these are not used in damaging ways.

Today most people in the developed world – and many beyond it – carry a highly sophisticated mapping device at all times. The rise of web-based mapping and GPS-enabled smartphones has radically changed the kinds of projects that critical scholars can undertake, while at the same time giving the tech sector an avalanche of data about our individual mobilities. It can surprise people when they discover just how many aspects of their life can be inferred from traces showing where they go and how long they spend there. Again, this puts a great responsibility onto critical scholars to protect participants when recording aspects of their mobility.

Final thoughts

There are, I think, four key lessons to draw from this book as a whole. The first may seem a little paradoxical given that the focus of this book is innovative methods. Nonetheless, for most of the projects that we consider, the kinds of old-fashioned methods familiar from our own disciplinary backgrounds will be the most appropriate. There is nothing inherently superior about using more innovative or complex methods where they are not needed to answer your research questions.

The second lesson is to acknowledge that playing around with a new method is not the easy option. Learning through doing involves a great deal of work as you get to grips with unfamiliar approaches and analytical frameworks. Working in this way is also unlikely to have an immediate pay-off in terms of the kind of outputs valued by managers working in the neoliberal university sector, so this approach does come with risk, particularly for those early in their academic careers.

The third lesson is that the key advantage of trying new methods is in broadening your perspective on the kinds of questions it is possible to ask and answer within a research project. Sometimes this will require a conversation or collaboration with someone who is an expert in the method that you have been trying out. Interdisciplinarity is never easy but it is an effective way to rapidly identify and explore new possibilities. Having already tried using a new technique can create a valuable point from which to have significantly more meaningful conversations with an expert from beyond your discipline.

The final lesson is about privilege. The tech sector is frequently and fairly criticised for its lack of diversity, with a culture that values the views and interests of wealthy white men. This has knock-on effects for the kinds of products it develops and the way that it manages the interests of its users. Given the global dominance of the sector, as critical scholars we have to be alert to how this privilege manifests itself in all aspects of our everyday lives. Nonetheless, it is too easy to simply dismiss new technologies as positivist and problematic. We must instead take a more nuanced view, engaging with the methodological opportunities that are presented while subjecting them to rigorous ethical scrutiny.

Working in the academic sector is a huge privilege in its own right. It gives a degree of freedom to explore new ideas, new techniques and ways of thinking that would be unimaginable to many working outside the academy. The approach of playing or experimenting with new methods that I have promoted in this book is a product of that privilege. For all of the hype, new technologies really can make a positive transformation to our everyday lives if employed carefully. As critical scholars we can, and perhaps should, take advantage of our position to try to make a positive difference in the world. An ethically informed approach to using methods that draw on new technologies is one way that we might do this.

References

Aarseth, E. (2001). Computer Game Studies, year one. *Game Studies*, 1, np.

Aarseth, E. (2003). Playing research: methodological approaches to game analysis. In Proceedings of the 5th international digital arts and culture conference, Melbourne, Australia, 19–25 May 2003, 1–7. http://heim.ifi.uio.no/~gisle/ifi/aarseth.pdf, accessed 27 January 2020.

Abedi, N., A. Bhaskar & E. Chung (2014). Tracking spatio-temporal movement of human in terms of space utilization using Media-Access-Control address data. *Applied Geography*, 51, 72–81.

Adler, J. (1979). *Artists in offices: an ethnography of an academic art scene.* Transaction Books, New Brunswick.

Ahas, R. (2011). Mobile positioning. In *Mobile methods*, eds. M. Büscher, J. Urry & K. Witchger, 183–199. Routledge, Abingdon.

Ahlers, M. (2013). TSA removes body scanners criticized as too revealing. https://edition.cnn.com/2013/05/29/travel/tsa-backscatter/index.html, accessed 16 August 2019.

Alexander, L. (2014). 'Gamers' don't have to be your audience. 'Gamers' are over. https://www.gamasutra.com/view/news/224400/Gamers_dont_have_to_be_your_audience_Gamers_are_over.php, accessed 11 February 2019.

Amati, M., J. Sita, E. Parmehr & C. McCarthy (2018). How eye-catching are natural features when walking through a park? Eye-tracking responses to videos of walks. *Urban Forestry & Urban Greening*, 31, 67–78.

Amoore, L. & A. Hall (2010). Border theatre: on the arts of security and resistance. *Cultural Geographies*, 17, 299–319.

Andreotti, A., P. Le Galès & F. J. Moreno Fuentes (2013). Transnational mobility and rootedness: the upper middle classes in European cities. *Global Networks*, 13, 41–59.

Arias, S. (2010). Rethinking space: an outsider's view of the spatial turn. *GeoJournal*, 75, 29–41.

Arnott, L. (2017). Mapping Metroid: narrative, space, and Other M. *Games and Culture*, 12, 3–27.

Arribas-Bel, D., K. Kourtit, P. Nijkamp & J. Steenbruggen (2015). Cyber cities: social media as a tool for understanding cities. *Applied Spatial Analysis and Policy*, 8, 231–247.

Aspinall, P., P. Mavros, R. Coyne & J. Roe (2015). The urban brain: analysing outdoor physical activity with mobile EEG. *British Journal of Sports Medicine*, 49, 272–276.

Bailenson, J. N., A. C. Beall, J. Loomis, J. Blascovich & M. Turk (2004). Transformed social interaction: decoupling representation from behavior and form in collaborative virtual environments. *Presence: Teleoperators and Virtual Environments*, 13, 428–441.

Bain, A. L. & F. Landau (2017). Artists, temporality, and the governance of collaborative place-making. *Urban Affairs Review,* 55, 405–427.

Baron, K. G., S. Abbott, N. Jao, N. Manalo & R. Mullen (2017). Orthosomnia: are some patients taking the quantified self too far? *Journal of Clinical Sleep Medicine,* 13, 351–354.

Barrass, S. & G. Kramer (1999). Using sonification. *Multimedia Systems,* 7, 23–31.

Beall, A. (2018). As Brexit looms, it's clear the tech to solve the Irish border problem is either untested or imaginary. https://www.wired.co.uk/article/irish-border-brexit-tech, accessed 13 August 2019.

Beer, D. (2017). The social power of algorithms. *Information, Communication & Society,* 20, 1–13.

Binchois Consort & A. Kirkman (2019). *Music for Saint Katherine of Alexandria.* Hyperion, London.

Birdsall, C. & D. Drozdzewski (2017). Capturing commemoration: using mobile recordings within memory research. *Mobile Media & Communication,* 6, 266–284.

Bishop, C. (2005). The social turn: collaboration and its discontents. *Artforum,* 44, 178.

Bissell, D. (2018). *Transit life: how commuting is transforming our cities.* MIT Press, Cambridge MA.

Blackwell, C., J. Birnholtz & C. Abbott (2014). Seeing and being seen: co-situation and impression formation using Grindr, a location-aware gay dating app. *New Media & Society,* 17, 1117–1136.

Boot, W. R., A. F. Kramer, D. J. Simons, M. Fabiani & G. Gratton (2008). The effects of video game playing on attention, memory, and executive control. *Acta Psychologica,* 129, 387–398.

Brandajs, F. & A. P. Russo (2019). Whose is that square? Cruise tourists' mobilities and negotiation for public space in Barcelona. *Applied Mobilities,* 1–25. doi:10.1080/23800127.2019.1576257

Brauneis, R. & E. P. Goodman (2018). Algorithmic transparency for the smart city. *Yale Journal of Law and Technology,* 20, 103–176.

Briscoe, G. & C. Mulligan (2014). Digital innovation: the hackathon phenomenon. *Creativeworks London Working Paper* No. 6, 1–13.

Brooke, S. (2018). Breaking gender code: hackathons, gender, and the social dynamics of competitive creation. In *Conference on human factors in computing systems,* 1–6. Montreal. http://hackathon-workshop-2018.com/Sian%20JM%20Brooke.pdf, accessed 27 January 2020.

Brooks, M. (1999). The porn pioneers. https://www.theguardian.com/technology/1999/sep/30/onlinesupplement, accessed 30 May 2019.

Brown, K. M., R. Dilley & K. Marshall (2008). Using a head-mounted video camera to understand social worlds and experiences. *Sociological Research Online,* 13, no pagination.

Brunsdon, C. & L. Comber (2015). *An introduction to R for spatial analysis and mapping.* Sage, London.

Bucher, T. (2012). Want to be on the top? Algorithmic power and the threat of invisibility on Facebook. *New Media & Society,* 14, 1164–1180.

Buck, U., S. Naether, B. Räss, C. Jackowski & M. J. Thali (2013). Accident or homicide: virtual crime scene reconstruction using 3D methods. *Forensic Science International,* 225, 75–84.

Burke, S. (2018). Your menstrual app is probably selling data about your body. https://www.vice.com/en_us/article/8xe4yz/menstrual-app-period-tracker-data-cyber-security, accessed 11 July 2019.

Büscher, M., J. Urry & K. Witchger (2011). *Mobile methods*. Routledge, Abingdon.

Calco, M. & A. Veeck (2015). The markathon: adapting the hackathon model for an introductory marketing class project. *Marketing Education Review*, 25, 33–38.

Campbell, A. (2017). The Google engineer's memo shows the stereotypes that keep women out of STEM. https://www.vox.com/new-money/2017/8/10/16118394/google-engineer-memo-stem, accessed 26 August 2019.

Cannella, G. S. & M. Koro-Ljungberg (2017). Neoliberalism in higher education: can we understand? Can we resist and survive? Can we become without neoliberalism? *Cultural Studies ↔ Critical Methodologies*, 17, 155–162.

Carpenter, L. L., T. T. Shattuck, A. R. Tyrka, T. D. Geracioti & L. H. Price (2011). Effect of childhood physical abuse on cortisol stress response. *Psychopharmacology*, 214, 367–375.

Carver, S. (2001). Public participation using web-based GIS. *Environment and Planning B-Planning & Design*, 28, 803–804.

Castro, J. C., M. Lalonde & D. Pariser (2016). Understanding the (im)mobilities of engaging at-risk youth through art and mobile media. *Studies in Art Education*, 57, 238–251.

Chang, E. (2018). *Brotopia: breaking up the boys' club of Silicon Valley*. Portfolio/Penguin, New York.

Chapman, A. (2016). It's hard to play in the trenches: World War I, collective memory and videogames. *Game Studies*, 16, np.

Cheney-Lippold, J. (2011). A new algorithmic identity: soft biopolitics and the modulation of control. *Theory, Culture & Society*, 28, 164–181.

Cieslak, M. (2016). Virtual reality to aid Auschwitz war trials of concentration camp guards. https://www.bbc.co.uk/news/technology-38026007, accessed 26 April 2019.

Cila, N., F. Tynan O'Mahony, E. Giaccardi, C. Speed, M. Caldwell & N. Rubens (2015). Listening to an everyday kettle: how can the data objects collect be useful for design research?. Proceedings of the 4th Participatory Innovation Conference, The Hague, 18–20 May 2015. http://pin-c.sdu.dk/assets/listening-to-an-everyday-kettle-how-can-the-data-objects-collect-be-useful-for-design-research.pdf, accessed 27 January 2020.

Clement, J. (2007). Visual influence on in-store buying decisions: an eye-track experiment on the visual influence of packaging design. *Journal of Marketing Management*, 23, 917–928.

Coemans, S. & K. Hannes (2017). Researchers under the spell of the arts: two decades of using arts-based methods in community-based inquiry with vulnerable populations. *Educational Research Review*, 22, 34–49.

Cope, M. & S. Elwood (2009). *Qualitative GIS: a mixed methods approach*. Sage, London.

Cosgrove, D. & S. Daniels (1988). *The iconography of landscape: essays on the symbolic representation, design and use of past environments*. Cambridge University Press, Cambridge.

Cote, A. C. (2017). "I can defend myself": women's strategies for coping with harassment while gaming online. *Games and Culture*, 12, 136–155.

Crampton, J. (2001). Maps as social constructions: power, communication and visualisation. *Progress in Human Geography*, 25, 235–252.

Crang, M. (2001). Rhythms of the city: temporalised space and motion. In *Timespace: geographies of temporality*, eds. J. May & N. Thrift, 187–207. Routledge, London.

Cresswell, T. (2006). *On the move: mobility in the modern western world*. Routledge, London.

Cresswell, T. (2010). Towards a politics of mobility. *Environment and Planning D: Society and Space*, 28, 17–31.

Csikszentmihalyi, M. (1990). *Flow: the psychology of optimal experience*. Harper & Row, New York.

David, G. & C. Cambre (2016). Screened intimacies: Tinder and the swipe logic. *Social Media + Society*, 2, 1–11.

DCMS (1998). *Creative industries mapping document*. DCMS, London.

DCMS (2010). Creative industries economic estimates - February 2010. http://weba rchive.nationalarchives.gov.uk/+/www.culture.gov.uk/reference_library/publications/ 6622.aspx, accessed 18 November 2010.

de Peuter, G. (2015). Online games and counterplay. In *The international encyclopedia of digital communication and society*, eds. R. Mansell & P. Ang. Hoboken, NJ: Wiley-Blackwell.

Debord, G. (2006 [1958]). Theory of the dérive. In *Situationist international anthology*, ed. K. Knabb. Bureau of Public Secrets, Berkeley.

Delfanti, A. (2012). Tweaking genes in your garage: biohacking between activism and entrepreneurship. In *Activist media and biopolitics: critical media interventions in the age of biopower*, eds. W. Sützl & T. Hug, 162–177. Innsbruck University Press, Innsbruck.

Derikx, S., M. de Reuver & M. Kroesen (2016). Can privacy concerns for insurance of connected cars be compensated? *Electronic Markets*, 26, 73–81.

DeSantis, A. D., E. M. Webb & S. M. Noar (2008). Illicit use of prescription ADHD medications on a college campus: a multimethodological approach. *Journal of American College Health*, 57, 315–324.

Deterding, S. (2017). The pyrrhic victory of game studies: assessing the past, present, and future of interdisciplinary game research. *Games and Culture*, 12, 521–543.

Diaz, J. L., C. Bil & A. Dyer (2017). Visual scan patterns of expert and cadet pilots in VFR landing. In *17th AIAA aviation technology, integration, and operations conference*, 1–11. American Institute of Aeronautics and Astronautics. https://doi.org/ 10.2514/6.2017-3777, accessed 27 January 2020.

digitalundivided (2018). ProjectDiane2018: the state of black women founders. http s://projectdiane.digitalundivided.com/, accessed 27 August 2019.

Disney, T., L. Warwick, H. Ferguson, J. Leigh, T. S. Cooner, L. Beddoe, P. Jones & T. Osborne (2019). "Isn't it funny the children that are further away we don't think about as much?": using GPS to explore the mobilities and geographies of social work and child protection practice. *Children and Youth Services Review*, 100, 39–49.

Dixon, S. (2019). Cybernetic-existentialism in performance art. *Leonardo*, 52, 247–254.

Dodge, M. (2013). Mappings. In *The Routledge handbook of mobilities*, eds. P. Adey, D. Bissell, K. Hannam, P. Merriman & M. Sheller, 517–533. Routledge, Abingdon.

Doerksen, M. D. (2017). Electromagnetism and the Nth sense: augmenting senses in the grinder subculture. *The Senses and Society*, 12, 344–349.

Duhaime-Ross, A. (2014). Apple promised an expansive health app, so why can't I track menstruation? https://www.theverge.com/2014/9/25/6844021/apple-promised-a n-expansive-health-app-so-why-cant-i-track, accessed 26 August 2019.

Duster, T. (2016). Ancestry testing and DNA: uses, limits–and caveat emptor. In *Genetics as social practice*, eds. B. Prainsack, S. Schicktanz & G. Werner-Felmayer, 59–72. Ashgate, Farnham.

Dwyer, T., C. Perkins, S. Redmond & J. Sita (2018*). Seeing into screens: eye tracking and the moving image*. Bloomsbury, New York.

Edmonds, E. & M. Leggett (2010). How artists fit into research processes. *Leonardo*, 43, 194–195.

Eghbal-Azar, K. & T. Widlok (2012). Potentials and limitations of mobile eye tracking in visitor studies. *Social Science Computer Review*, 31, 103–118.

Ekman, M. & A. Widholm (2017). Political communication in an age of visual connectivity: exploring Instagram practices among Swedish politicians. *Northern Lights: Film & Media Studies Yearbook*, 15, 15–32.

Elson, M. & C. J. Ferguson (2014). Twenty-five years of research on violence in digital games and aggression: empirical evidence, perspectives, and a debate gone astray. *European Psychologist*, 19, 33–46.

Elwood, S. (2006). Negotiating knowledge production: the everyday inclusions, exclusions, and contradictions of Participatory GIS research. *The Professional Geographer*, 58, 197–208.

Emo, B., K. Al-Sayed & T. Varoudis (2016). Design, cognition and behaviour: usability in the built environment. *International Journal of Design Creativity and Innovation*, 4, 63–66.

Entertainment Software Association (2018). *Essential facts about the computer and video game industry: 2018 sales, demographic and usage data*. ESA, Washington DC.

European Commission (2014). Factsheet on the "Right to be Forgotten" ruling (C-131/12). https://www.inforights.im/media/1186/cl_eu_commission_factsheet_right_to_be-forgotten.pdf, accessed 5 June 2018.

European Commission (2018). A new era for data protection in the EU. What changes after May 2018? https://ec.europa.eu/commission/sites/beta-political/files/data-protection-factsheet-changes_en.pdf, accessed 5 June 2018.

Evans, D. (2001). *Emotion: the science of sentiment*. Oxford University Press, Oxford.

Evans, J. & P. Jones (2008). Towards Lefebvrian socio-nature? A film about rhythm, nature and science. *Geography Compass*, 2, 659–670.

Evans, J. & P. Jones (2011). The walking interview: methodology, mobility and place. *Applied Geography*, 31, 849–858.

Fardouly, J., B. K. Willburger & L. R. Vartanian (2017). Instagram use and young women's body image concerns and self-objectification: testing mediational pathways. *New Media & Society*, 20, 1380–1395.

Farnsworth, A. (2016). Self-sampling HPV testing versus mainstream cervical screening and HPV testing. *Medical Journal of Australia*, 204, 171. e1.

Félix, P., J. Figueiredo, C. P. Santos & J. C. Moreno (2017). Adaptive real-time tool for human gait event detection using a wearable gyroscope. In *Human-centric robotics*, eds. M. Silva, G. Virk, M. Tokhi, B. Malheiro, P. Guedes & P. Ferreira, 653–660. World Scientific, Singapore.

Ferguson, E. L. (2015). Facial identification of children: a test of automated facial recognition and manual facial comparison techniques on juvenile face images. Unpublished PhD Thesis, University of Dundee.

Field, J. (1950). *On not being able to paint*. Heinemann, London.

Fisher, P. & D. Unwin (2002). *Virtual reality in geography*. Taylor & Francis, London.

Fleming, C. (2002). Performance as guerrilla ontology: the case of Stelarc. *Body & Society*, 8, 95–109.

Florida, R. (2002). *The rise of the creative class and how it's transforming work, leisure, community and everyday life*. Basic Books, New York.

Fokkert, M. J., P. R. van Dijk, M. A. Edens, S. Abbes, D. de Jong, R. J. Slingerland & H. J. G. Bilo (2017). Performance of the FreeStyle Libre Flash glucose monitoring system in patients with type 1 and 2 diabetes mellitus. *BMJ Open Diabetes Research & Care*, 5, e000320.

Foster, K. & H. Lorimer (2007). Some reflections on art-geography as collaboration. *Cultural Geographies*, 14, 425–432.

Foucault, M. (1977). *Discipline and punish: the birth of the prison*. Allen Lane, London.

Frasca, G. (2003). Ludologists love stories, too: notes from a debate that never took place. In *Proceedings of DiGRA 2003: level up*, eds. M. Copier & J. Raessens, 92–99. University of Utrecht, Utrecht.

Freeman, J. C. & M. Sheller (2015). Editors' statement: hybrid space and digital public art. *Public Art Dialogue*, 5, 1–8.

Gale, R. (2013). Religious residential segregation and internal migration: the British Muslim case. *Environment and Planning A: Economy and Space*, 45, 872–891.

Germann Molz, J. & C. M. Paris (2015). The social affordances of flashpacking: exploring the mobility nexus of travel and communication. *Mobilities*, 10, 173–192.

Ghaffary, S. (2019). Political tension at Google is only getting worse. https://www.vox.com/recode/2019/8/2/20751822/google-employee-dissent-james-damore-cernekee-conservatives-bias, accessed 28 August 2019.

Gibson, W. (2007). *Spook country*. Viking Press, New York.

Gilleade, K., A. Dix & J. Allanson (2005). Affective videogames and modes of affective gaming: assist me, challenge me, emote me. Proceedings of DiGRA 2005 conference: changing views – worlds in play, 1–7. www.digra.org/digital-library/publications/affective-videogames-and-modes-of-affective-gaming-assist-me-challenge-me-emote-me/, accessed 27 January 2020.

Gioia, T. (2015). San Francisco during the great food awakening. *Virginia Quarterly Review*, 91, 250–255.

Gitelman, L. (2013). *"Raw data" is an oxymoron*. MIT Press, Cambridge, MA.

Glouftsios, G. (2018). Governing circulation through technology within EU border security practice-networks. *Mobilities*, 13, 185–199.

Goldstaub, T. (2017). The dangers of tech-bro AI. *MIT Technology Review*, November/December, n.p.

Good, O. (2019). Hololens headsets for the Army blasted by Microsoft workers. https://www.polygon.com/2019/2/23/18237553/hololens-army-contract-microsoft-protest, accessed 30 May 2019.

Goodall, J. (1999). An order of pure decision: un-natural selection in the work of Stelarc and Orlan. *Body & Society*, 5, 149–170.

Goodchild, M. (2007). Citizens as sensors: the world of volunteered geography. *GeoJournal*, 69, 211–221.

Gordon, E. & G. Koo (2008). Placeworlds: using virtual worlds to foster civic engagement. *Space and Culture*, 11, 204–221.

Gössling, S. & I. Stavrinidi (2016). Social networking, mobilities, and the rise of liquid identities. *Mobilities*, 11, 723–743.

Graham, D. J., J. L. Orquin & V. H. M. Visschers (2012). Eye tracking and nutrition label use: a review of the literature and recommendations for label enhancement. *Food Policy*, 37, 378–382.

Gulson, K. & C. Symes (2007). *Spatial theories of education: policy and geography matters*. Routledge, London.

Hadfield-Hill, S. & C. Zara (2018). Being participatory through the use of app-based research tools. In *Being participatory: researching with children and young people: co-constructing knowledge using creative techniques*, eds. I. Coyne & B. Carter, 147–169. Springer International Publishing, Cham.

Haklay, M. (2013). Neogeography and the delusion of democratisation. *Environment and Planning A*, 45, 55–69.

Haraway, D. (1985). Manifesto for cyborgs: science, technology, and socialist feminism in the 1980s. *Socialist Review*, 80, 65–108.

Harley, J. B. (1989). Deconstructing the map. *Cartographica*, 26, 1–20.

Harrowell, E., T. Davies & T. Disney (2018). Making space for failure in geographic research. *The Professional Geographer*, 70, 230–238.

Hayes, A. & J. Hunter (2012). Why is publication of negative clinical trial data important? *British Journal of Pharmacology*, 167, 1395–1397.

Heikenfeld, J. (2016). Non-invasive analyte access and sensing through eccrine sweat: challenges and outlook circa 2016. *Electroanalysis*, 28, 1242–1249.

Heim, M. R. (2017). Virtual reality wave 3. In *Boundaries of self and reality online*, eds. J. Gackenbach & J. Bown, 261–277. Academic Press, San Diego.

Helmore, E. (2019). Hundreds of Google employees urge company to resist support for Ice. https://www.theguardian.com/technology/2019/aug/16/hundreds-of-google-employees-urge-company-to-resist-support-for-ice, accessed 28 August 2019.

Herborn, K. A., J. L. Graves, P. Jerem, N. P. Evans, R. Nager, D. J. McCafferty & D. E. F. McKeegan (2015). Skin temperature reveals the intensity of acute stress. *Physiology & Behavior*, 152, 225–230.

Hern, A. (2018). Strava suggests military users 'opt out' of heatmap as row deepens. https://www.theguardian.com/technology/2018/jan/29/strava-secret-army-base-locations-heatmap-public-users-military-ban, accessed 3 July 2018.

Hernandez, P. (2019). Trump and GOP blame recent mass shootings on video games, internet. https://www.polygon.com/2019/8/5/20754784/el-paso-dayton-mass-shootings-trump-video-games, accessed 26 August 2019.

Hess, M., V. Petrovic, D. Meyer, D. Rissolo & F. Kuester (2015). Fusion of multimodal three-dimensional data for comprehensive digital documentation of cultural heritage sites. In Proceedings of 2015 digital heritage conference, 595–602. https://ieeexplore.ieee.org/document/7419578, accessed 27 January 2020.

Hill, R. (2018). London's Met Police: we won't use facial recognition at Notting Hill Carnival. https://www.theregister.co.uk/2018/05/24/met_police_wont_use_facial_recognition_tech_at_notting_hill_this_year/, accessed 8 June 2018.

Hillier, B. & J. Hanson (1984). *The social logic of space*. Cambridge University Press, Cambridge.

Hillis, K. (1996). A geography of the eye: the technologies of virtual reality. In *Cultures of the internet: virual spaces, real histories, living bodies*, ed. R. Shields, 70–98. Sage, London.

Holley, D., J. Jain & G. Lyons (2008). Understanding business travel time and its place in the working day. *Time & Society*, 17, 27–46.

Holloway-Attaway, L. (2013). Performing materialities: exploring mixed media reality and Moby-Dick. *Convergence*, 20, 55–68.

Holmqvist, K., M. Nyström, R. Anderson, R. Dewhurst, H. Jarodzka & J. van de Weijer (2011). *Eye tracking: a comprehensive guide to methods and measures*. Oxford University Press, Oxford.

Holton, M. (2019). Walking with technology: understanding mobility-technology assemblages. *Mobilities*, 14, 435–451.

Humphreys, S. (2019). On being a feminist in Games Studies. *Games and Culture*, 14, 825–842.

Hurley, R., D. Hutcherson, C. Tonkin, S. Dailey & J. Rice (2015). Measuring physiological arousal towards packaging: tracking electrodermal activity within the consumer shopping environment. *Journal of Applied Packaging Research*, 7, 76–90.

Husić-Mehmedović, M., I. Omeragić, Z. Batagelj & T. Kolar (2017). Seeing is not necessarily liking: advancing research on package design with eye-tracking. *Journal of Business Research*, 80, 145–154.

Idhe, D. (1990). *Technology and the lifeworld: from the garden to the earth*. Indiana University Press, Bloomington.

Ilieva, I. P., C. J. Hook & M. J. Farah (2015). Prescription stimulants' effects on healthy inhibitory control, working memory, and episodic memory: a meta-analysis. *Journal of Cognitive Neuroscience*, 27, 1069–1089.

Isaac, M. (2017). Uber founder Travis Kalanick resigns as C.E.O. https://www.nytimes.com/2017/06/21/technology/uber-ceo-travis-kalanick.html, accessed 26 August 2019.

Isakjee, A. (2019). Engineering cohesion: a reflection on academic practice in a community-based setting. In *Cultural intermediaries connecting communities: revisiting approaches to cultural engagement*, eds. P. Jones, B. Perry & P. Long, 169–180. Policy Press, Bristol.

Isakjee, A. & C. Allen (2013). 'A catastrophic lack of inquisitiveness': a critical study of the impact and narrative of the Project Champion surveillance project in Birmingham. *Ethnicities*, 13, 751–770.

Ivancheva, M. P. (2015). The age of precarity and the new challenges to the academic profession. *Studia Universitatis Babes-Bolyai-Studia Europaea*, 60, 39–48.

Ivory, J. D. (2006). Still a man's game: gender representation in online reviews of video games. *Mass Communication and Society*, 9, 103–114.

Jam, C. (2019). Intervention: force deep. In *Cultural intermediaries connecting communities: revisiting approaches to cultural engagement*, eds. P. Jones, B. Perry & P. Long, 181–183. Policy Press, Bristol.

Johnstone, J. A., P. A. Ford, G. Hughes, T. Watson, A. C. S. Mitchell & A. T. Garrett (2012). Field based reliability and validity of the Bioharness™ multivariable monitoring device. *Journal of Sports Science & Medicine*, 11, 643–652.

Jones, P. (2005). Performing the city: a body and a bicycle take on Birmingham, UK. *Social and Cultural Geography*, 6, 813–830.

Jones, P. (2014). Performing sustainable transport: an artistic RIDE across the city. *Cultural Geographies*, 21, 287–292.

Jones, P. & J. Evans (2011). Creativity and project management: a comic. *ACME: An International e-Journal for Critical Geographies*, 10, 585–632.

Jones, P. & J. Evans (2012a). Rescue geography: place making, affect and regeneration. *Urban Studies*, 49, 2315–2330.

Jones, P. & J. Evans (2012b). The spatial transcript: analysing mobilities through qualitative GIS. *Area*, 44, 92–99.

Jones, P. & C. Jam (2016). Creating ambiances, co-constructing place: a poetic transect across the city. *Area*, 48, 317–324.

Jones, P. & T. Osborne (2019). Analysing virtual landscapes using postmemory. *Social & Cultural Geography*, 1–21. doi:10.1080/14649365.2018.1474378

Jones, P. & N. Macdonald (2007). Getting it wrong first time: building an interdisciplinary research relationship. *Area*, 39, 490–498.

Jones, P., R. Drury & J. McBeath (2011). Using GPS-enabled mobile computing to augment qualitative interviewing: two case studies. *Field Methods*, 23, 173–187.

Jones, P., A. Isakjee, C. Jam, C. Lorne & S. Warren (2017). Urban landscapes and the atmosphere of place: exploring subjective experience in the study of urban form. *Urban Morphology*, 21, 29–40.

Jones, P., A. Layard, C. Speed & C. Lorne (2015). MapLocal: use of smartphones for crowdsourced planning. *Planning Practice & Research*, 30, 322–336.

Jones, P., P. Long & B. Perry (2019). Conclusion: where next for cultural intermediation? In *Cultural intermediaries connecting communities: revisiting approaches to cultural engagement*, eds. P. Jones, B. Perry & P. Long, 223–232. Policy Press, Bristol.

Kang, C. (2019). F.T.C. approves Facebook fine of about $5 billion. https://www.nytimes.com/2019/07/12/technology/facebook-ftc-fine.html, accessed 23 September 2019.

Kaplan, S. (1995). The restorative benefits of nature: toward an integrative framework. *Journal of Environmental Psychology*, 15, 169–182.

Katz, M. (2018). Augmented Reality is transforming museums. https://www.wired.com/story/augmented-reality-art-museums/, accessed 31 May 2019.

Kaufman, E. (2016). Policing mobilities through bio-spatial profiling in New York city. *Political Geography*, 55, 72–81.

Kemp, M. (2011). Culture: artists in the lab. *Nature*, 477, 278–279.

Khakurel, J., H. Melkas & J. Porras (2018). Tapping into the wearable device revolution in the work environment: a systematic review. *Information Technology & People*, 31, 791–818.

Kirkman, A. & P. Weller (2020 in press). English alabaster images as recipients of music in the long fifteenth century: English sacred traditions in a European perspective. In *English alabaster carvings and their cultural contexts*, ed. Z. Murat. Boydell & Brewer, Martlesham.

Kitchin, R. (2014). *The data revolution: big data, open data, data infrastructures and their consequences.* Sage, London.

Kitchin, R. & G. McArdle (2018). Urban data and city dashboards: six key issues. In *Data and the city*, eds. R. Kitchin, T. Lauriault & G. McArdle, 111–126. Routledge, Abingdon.

Klein, C. (2010). Philosophical issues in neuroimaging. *Philosophy Compass*, 5, 186–198.

Kljun, M., K. Čopič, Pucihar & P. Coulton (2018). User engagement continuum: art engagement and exploration with Augmented Reality. In *Augmented Reality art: from an emerging technology to a novel creative medium*. Second edition, ed. V. Geroimenko, 329–342. Springer, Cham.

Kolla, B. P., S. Mansukhani & M. P. Mansukhani (2016). Consumer sleep tracking devices: a review of mechanisms, validity and utility. *Expert Review of Medical Devices*, 13, 497–506.

Koski, K. & J. Holst (2017). Exploring vaccine hesitancy through an artist–scientist collaboration. *Journal of Bioethical Inquiry*, 14, 411–426.

Krauzlis, R. J. (2004). Recasting the smooth pursuit eye movement system. *Journal of Neurophysiology*, 91, 591–603.

Kristian, M. & R. Brady (2019). Walking through, going along and scrolling back. *Nordicom Review*, 40, 95–109.

Kuhn, T. (1970). *The structure of scientific revolutions*. Second edition. University of Chicago Press, Chicago.

Kusenbach, M. (2003). Street phenomenology: the go-along as ethnographic research tool. *Ethnography*, 4, 455–485.

Kwan, M.-P. & G. Ding (2008). Geo-narrative: extending Geographic Information Systems for narrative analysis in qualitative and mixed-method research. *The Professional Geographer*, 60, 443–465.

Lartey, J. (2018). Donald Trump rains insults on Elizabeth Warren after DNA test. https://www.theguardian.com/us-news/2018/oct/16/donald-trump-elizabeth-wa rren-dna-test, accessed 22 July 2019.

Laurier, E. & C. Philo (2003). The region in the boot: mobilising lone subjects and multiple objects. *Environment and Planning D*, 21, 85–106.

Laurier, E. & C. Philo (2006). Possible geographies: a passing encounter in a café. *Area*, 38, 353–363.

Lazer, D. (2015). The rise of the social algorithm. *Science*, 348, 1090–1091.

LeBlanc, A. G. & J.-P. Chaput (2017). Pokémon Go: a game changer for the physical inactivity crisis? *Preventive Medicine*, 101, 235–237.

Lee, K., A. Agrawal & A. Choudhary (2013). Real-time disease surveillance using Twitter data: demonstration on flu and cancer.In *Proceedings of the 19th ACM SIGKDD international conference on knowledge discovery and data mining*, 1474–1477. ACM, Chicago, Illinois.

Levy, I. (2016). The smart security behind the GB Smart Metering System. https://www.ncsc.gov.uk/articles/smart-security-behind-gb-smart-metering-system, accessed 10 August 2018.

Lewinski, P., J. Trzaskowski & J. Luzak (2016). Face and emotion recognition on commercial property under EU data protection law. *Psychology & Marketing*, 33, 729–746.

Lewis, A. (2014). Assassin's Creed Unity - where's your accent. http://blog.ubi.com/en-GB/assassins-creed-unity-accents/, accessed 26 April 2019.

Liberati, N. (2017). Teledildonics and new ways of "being in touch": a phenomenological analysis of the use of haptic devices for intimate relations. *Science and Engineering Ethics*, 23, 801–823.

Lloyd, M. (2019). The non-looks of the mobile world: a video-based study of interactional adaptation in cycle-lanes. *Mobilities*, 14, 500–523.

Longan, M. W. (2008). Playing with landscape: social process and spatial form in video games. *Aether: The Journal of Media Geography*, 11, 23–40.

Losse, K. (2012). *The boy kings: a journey into the heart of the social network*. Free Press, New York.

Losse, K. (2016). The art of failing upward. https://www.nytimes.com/2016/03/06/opinion/sunday/the-art-of-failing-upward.html, accessed 27 August 2019.

Lovelace, C. (1995). Orlan: offensive acts. *Performing Arts Journal*, 17, 13–25.

Lowensohn, J. (2014). Apple's first diversity report shows company to be mostly male, white. https://www.theverge.com/2014/8/12/5949453/no-surprise-apple-is-very-white-very-male, accessed 26 August 2019.

Lukyanov, E. (2019). Nietzschean superhuman evolution in decentralized technological era. In *The transhumanism handbook*, ed. N. Lee, 651–657. Springer International Publishing, Cham.

Lupton, D. (2015). Quantified sex: a critical analysis of sexual and reproductive self-tracking using apps. *Culture, Health & Sexuality*, 17, 440–453.

Lupton, D. (2016). *The quantified self: a sociology of self-tracking*. Polity, Cambridge.

Lyons, G. & J. Urry (2005). Travel time use in the information age. *Transportation Research Part A: Policy and Practice*, 39, 257–276.

Malatino, H. (2017). Biohacking gender. *Angelaki*, 22, 179–190.

Marsh, S. (2018). Extreme biohacking: the tech guru who spent $250,000 trying to live for ever. https://www.theguardian.com/science/2018/sep/21/extreme-biohacking-tech-guru-who-spent-250000-trying-to-live-for-ever-serge-faguet, accessed 10 July 2019.

Martin, A. (2015). Download Festival face scan: you're right to be annoyed, said UK surveillance commish. https://www.theregister.co.uk/2015/07/13/sneaky_use_of_facial_recognition_at_download_rightly_caused_outcry_according_to_blightys_surveillance_commish/, accessed 8 June 2018.

McCall, C. (2018). Brexit, bordering and bodies on the island of Ireland. *Ethnopolitics*, 17, 292–305.

McRobbie, A. (2004). Post-feminism and popular culture. *Feminist Media Studies*, 4, 255–264.

Mehta, D., B. Dedhia, B. Kothari & U. Joshi (2017). Automated attendance monitoring system using facial recognition. *International Journal of Engineering Science*, 7, 11707–11709.

Michael, K. & M. Michael (2012). Implementing 'namebers' using microchip implants: the black box beneath the skin. In *This pervasive day: the potential and perils of pervasive computing*, ed. J. Pitt, 163–206. Imperial College Press, London.

Miller, C. (2006). A beast in the field: the Google Maps mashup as GIS/2. *Cartographica*, 41, 187–199.

Miller, P. (1998). The engineer as catalyst: Billy Klüver on working with artists. *IEEE Spectrum*, 35, 20–29.

Mok, G. T. K. & B. H.-Y. Chung (2017). AB119. Computer-aided facial recognition of Chinese individuals with 22q11.2 deletion-algorithm training using NIH atlas of human malformation syndromes from diverse population. *Annals of Translational Medicine*, 5, AB119.

Moore, P. (2017). *The quantified self in precarity: work, technology and what counts*. Routledge, London.

Moore, P. & A. Robinson (2015). The quantified self: what counts in the neoliberal workplace. *New Media & Society*, 18, 2774–2792.

Mortensen, T. E. (2018). Anger, fear, and games: the long event of #GamerGate. *Games and Culture*, 13, 787–806.

Mukherjee, S. (2017). *Videogames and postcolonialism: empire plays back*. Palgrave Macmillan, London.

Mukherjee, S. (2018). Playing subaltern: video games and postcolonialism. *Games and Culture*, 13, 504–520.

Muscat, A., W. Goddard, J. Duckworth & J. Holopainen (2016). First-person walkers: understanding the walker experience through four design themes. In Proceedings of 1st international joint conference of DiGRA and FDG, Dundee, UK, 1–6 August 2016, 1–15. www.digra.org/digital-library/publications/first-person-walkers-the-walker-experience-u-tnhdroerusgtha-nfdoiunrg-design-themes/, accessed 27 January 2020.

Nafus, D. (2016). Introduction. In *Quantified: biosensing technologies in everyday life*, ed. D. Nafus, ix–xxxi. MIT Press, Cambridge MA.

Nagata, J. M., T. A. Brown, J. M. Lavender & S. B. Murray (2019). Emerging trends in eating disorders among adolescent boys: muscles, macronutrients, and biohacking. *The Lancet Child & Adolescent Health*, 3, 444–445.

Nash, D. B. (2010). Beware biohacking? *Biotechnology Healthcare*, 7, 7.

Neff, G. & D. Nafus (2016). *Self-tracking*. MIT Press, Cambridge MA.

Nellis, M. (2009). 24/7/365: mobility, locatability and the satellite tracking of offenders. In *Technologies of insecurity: the surveillance of everyday life*, eds. F. Aas, H. Grundhus & H. Lomell, 105–124. Routledge-Cavendish, Abingdon.

Newzoo (2018). 2018 global games market. https://newzoo.com/key-numbers/, accessed 23 January 2019.

Nixon, P. (2017). Hell yes! Playing away, teledildonics and the future of sex. In *Sex in the digital age*, eds. P. Nixon & I. Düsterhöft, 201–212. Routledge, London.

Noble, S. U. (2018). *Algorithms of oppression: how search engines reinforce racism.* NYU Press, New York.

Nold, C. (2009). Greenwich emotion map. In *Emotional cartography: technologies of the self*, ed. C. Nold, 63–67. http://emotionalcartography.net/EmotionalCartography. pdf, accessed 12 February 2010.

O'Rourke, K. (2013). *Walking and mapping: artists as cartographers.* MIT Press, Cambridge MA.

Obermeyer, N. (1998). The evolution of public participation GIS. *Cartography and Geographical Information Science*, 25, 65–66.

Ochsner, A. (2019). Reasons why: examining the experience of women in games 140 characters at a time. *Games and Culture*, 14, 523–542.

Osborne, T. & P. I. Jones (2017). Biosensing and geography: a mixed methods approach. *Applied Geography*, 87, 160–169.

Osborne, T., E. Warner, P. I. Jones & B. Resch (2019). Performing social media: artistic approaches to analyzing big data. *GeoHumanities*, 5, 282–294.

Page, C. (2018). Security bods slam 'hackable' smart meters as firms prepare for SMETS 2 rollout. https://www.theinquirer.net/inquirer/news/3026904/security-bods-slam -hackable-smart-meters-as-energy-firms-prepare-for-smets-2-rollout, accessed 10 August 2018.

Paglen, T. (2009). Experimental geography: from cultural production to the production of space. In *Experimental geography: landscape hacking, cartography, and radical urbanism*, eds. N. Thompson & Independent Curators, np. Melville House Publishing, Brooklyn.

paintersflat (2007). locative media land art. https://www.youtube.com/watch?v=lMp AeuJd6YQ, accessed 15 August 2019.

Pallasmaa, J. (2005). *The eyes of the skin: architecture and the senses.* John Wiley, Chichester.

Papagiannis, H. (2014). Working towards defining an aesthetics of augmented reality: a medium in transition. *Convergence*, 20, 33–40.

Parks, C. L. & K. L. Monson (2017). Automated facial recognition of computed tomography-derived facial images: patient privacy implications. *Journal of Digital Imaging*, 30, 204–214.

Parks, C. L. & K. L. Monson (2018). Automated facial recognition of manually generated clay facial approximations: potential application in unidentified persons data repositories. *Forensic Science International*, 282, 133–136.

Parlak, O., S. T. Keene, A. Marais, V. F. Curto & A. Salleo (2018). Molecularly selective nanoporous membrane-based wearable organic electrochemical device for noninvasive cortisol sensing. *Science Advances,* 4, eaar2904. doi:10.1126/sciadv. aar2904.

Patney, A., M. Salvi, J. Kim, A. Kaplanyan, C. Wyman, N. Benty, D. Luebke & A. Lefohn (2016). Towards foveated rendering for gaze-tracked virtual reality. *ACM Transactions on Graphics,* 35, 1–12.

Paul, C. (2018). *The toxic meritocracy of video games: why gaming culture is the worst.* University of Minnesota Press, Minneapolis.

Perez, S. (2015). Apple stops ignoring women's health with iOS 9 HealthKit update, now featuring period tracking. https://techcrunch.com/2015/06/09/apple-stops-ignor ing-womens-health-with-ios-9-healthkit-update-now-featuring-period-tracking/, accessed 26 August 2019.

Picard, R. (1997). *Affective computing.* MIT Press, Cambridge MA.

Pieters, R. & M. Wedel (2008). A review of eye-tracking research in marketing. In *Review of marketing research: volume 4,* ed. N. Malhotra, 123–147. Emerald Publishing, Bingley.

Pinchbeck, D. (2008). Dear Esther: an interactive ghost story built using the source engine. In *Interactive storytelling. Proceedings of the first joint international conference on interactive digital storytelling, ICIDS 2008 Erfurt, Germany, November 26–29, 2008,* eds. U. Spierling & N. Szilas, 51–54. Springer, Berlin, Heidelberg.

Pinney, A. (2005). Ethics, agency, and desire in two strip clubs: a view from both sides of the gaze. *Qualitative Inquiry,* 11, 716–723.

Porter, G., K. Hampshire, A. de Lannoy, N. Gungguluza, M. Mashiri & A. Bango (2018). Exploring the intersection between physical and virtual mobilities in urban South Africa: reflections from two youth-centred studies. In *Urban mobilities in the global south,* eds. T. P. Uteng & K. Lucas, 61–77. Routledge, Abingdon.

Porter, J. (2019). Group dating app found leaking basically everything about its users worldwide. https://www.theverge.com/2019/8/9/20798290/3fun-data-breach-secur ity-cybersecurity-group-dating-app, accessed 15 August 2019.

Portman, M. E., A. Natapov & D. Fisher-Gewirtzman (2015). To go where no man has gone before: virtual reality in architecture, landscape architecture and environmental planning. *Computers, Environment and Urban Systems,* 54, 376–384.

Posada-Quintero, H. F., N. Reljin, C. Mills, I. Mills, J. P. Florian, J. L. VanHeest & K. H. Chon (2018). Time-varying analysis of electrodermal activity during exercise. *PloS one,* 13, e198328–e198328.

Posner, J., J. A. Russell & B. S. Peterson (2005). The circumplex model of affect: an integrative approach to affective neuroscience, cognitive development, and psychopathology. *Development and Psychopathology,* 17, 715–734.

Prasad, R. (2018). Think tank: is Wi-Fi tracking becoming a thing of the past?https://wwd. com/business-news/technology/wifi-tracking-a-thing-of-the-past-think-tank-1202772454/, accessed 11 July 2019.

Pred, A. (1977). The choreography of existence: comments on Hägerstrand's time-geography and its usefulness. *Economic Geography,* 53, 207–221.

Pschetz, L., E. Tallyn, R. Gianni & C. Speed (2017). Bitbarista: exploring perceptions of data transactions in the Internet of Things. *Proceedings of the 2017 CHI conference on human factors in computing systems,* 2964–2975. https://doi.org/10.1145/ 3025453.3025878, accessed 27 January 2020.

Public Health England (2018). Brisk walking and physical inactivity in 40 to 60 year olds. https://www.gov.uk/government/publications/brisk-walking-and-physical-inactivity-in-

40-to-60-year-olds/brisk-walking-and-physical-inactivity-in-40-to-60-year-olds, accessed 8 July 2019.

Rapp, D. (2014). Sex in the cinema: war, moral panic, and the British film industry, 1906–1918. *Albion*, 34, 422–451.

Reeves, A. (2014). Neither class nor status: arts participation and the social strata. *Sociology*, 49, 624–642.

Reeves, S., B. Brown & E. Laurier (2009). Experts at play: understanding skilled expertise. *Games and Culture*, 4, 205–227.

Rembar, C. (1969). *The end of obscenity: the trials of Lady Chatterley, Tropic of Cancer & Fanny Hill*. André Deutsch, London.

Resch, B., A. Summa, G. Sagl, P. Zeile & J.-P. Exner (2015). Urban emotions: geosemantic emotion extraction from technical sensors, human sensors and crowdsourced data. In *Progress in location-based services*, eds. G. Gartner & H. Huang, 199–212. Springer, New York.

Richardson, A. E., M. E. Powers & L. G. Bousquet (2011). Video game experience predicts virtual, but not real navigation performance. *Computers in Human Behavior*, 27, 552–560.

Roberts, H., J. Sadler & L. Chapman (2019). The value of Twitter data for determining the emotional responses of people to urban green spaces: a case study and critical evaluation. *Urban Studies*, 56, 818–835.

Robertson, A. (2019). The Valve Index might have the most fun VR controllers I've ever tried. https://www.theverge.com/2019/5/28/18639084/valve-index-steamvr-headset-knuckles-controllers-preview, accessed 30 May 2019.

Rodden, J. (1987). The spectre of Der Gross Bruder: George Orwell's reputation in West Germany. *The German Quarterly*, 60, 530–547.

Rollet, C. (2018). The odd reality of life under China's all-seeing credit score system. https://www.wired.co.uk/article/china-social-credit, accessed 10 August 2018.

Romano Bergstrom, J. C., E. L. Olmsted-Hawala & M. E. Jans (2013). Age-related differences in eye tracking and usability performance: website usability for older adults. *International Journal of Human–Computer Interaction*, 29, 541–548.

Romano Bergstrom, J. C. & A. Schall (2014). *Eye tracking in user experience design*. Morgan Kaufman, Waltham, MA.

Rosch, J. L. & J. J. Vogel-Walcutt (2013). A review of eye-tracking applications as tools for training. *Cognition, Technology & Work*, 15, 313–327.

Rose, G. (1992). Geography as science of observation: the landscape, the gaze and masculinity. In *Nature and science: essays in the history of geographical knowledge*, eds. F. Driver & G. Rose, 8–18. Historical Geography Research Series, London.

Rose, G. (1993). *Feminism and geography: the limits of geographical knowledge*. Polity Press, Cambridge.

Rose, G. (2000). Practising photography: an archive, a study, some photographs and a researcher. *Journal of Historical Geography*, 26, 555–571.

Rose, G. (2010). *Doing family photography: the domestic, the public and the politics of sentiment*. Ashgate, Farnham.

Rotter, P., B. Daskala & R. Compano (2008). RFID implants: opportunities and challenges for identifying people. *IEEE Technology and Society Magazine*, 27, 24–32.

Rubens, T. (2018). Drug-smuggling drones: how prisons are responding to the airborne security threat. https://www.ifsecglobal.com/drones/drug-smuggling-drones-prisons-airborne-security-threat/, accessed 16 August 2019.

Rubin, P. (2018). Coming attractions: the rise of VR porn. https://www.wired.com/story/coming-attractions-the-rise-of-vr-porn/, accessed 30 May 2019.

Russell, J. A. (1980). A circumplex model of affect. *Journal of Personality and Social Psychology*, 39, 1161–1178.

Sainio, J., J. Westerholm & J. Oksanen (2015). Generating heat maps of popular routes online from massive mobile sports tracking application data in milliseconds while respecting privacy. *ISPRS International Journal of Geo-Information*, 4, 1813–1826.

Schlesinger, A., W. K. Edwards & R. E. Grinter (2017). Intersectional HCI: engaging identity through gender, race, and class. Proceedings of the 2017 CHI conference on human factors in computing systems, Denver, Colorado, USA, 5412–5427. https://doi.org/10.1145/3025453.3025766, accessed 27 January 2020.

Schneider, A. (2013). Contested grounds: fieldwork collaborations with artists in Corrientes, Argentina. *Critical Arts*, 27, 511–530.

Schumann, C., N. D. Bowman & D. Schultheiss (2016). The quality of video games: subjective quality assessments as predictors of self-reported presence in first-person shooter and role-playing games. *Journal of Broadcasting & Electronic Media*, 60, 547–566.

Scobie, A. (1990). *Hitler's state architecture: the impact of classical antiquity.* Pennsylvania University Press, London.

Seppala, T. (2016). Crafting the alorithmic soundtrack of 'No Man's Sky'. https://www.engadget.com/2016/08/11/no-mans-sky-soundtrack-65daysofstatic-interview, accessed 23 April 2019.

Shanken, E. A. (2005). Artists in industry and the academy: collaborative research, interdisciplinary scholarship and the creation and interpretation of hybrid forms. *Leonardo*, 38, 415–418.

Sharma, S. (2014). *In the meantime: temporality and cultural politics.* Duke University Press, Durham NC.

Sheller, M. (2017). From spatial turn to mobilities turn. *Current Sociology*, 65, 623–639.

Sheller, M. & J. Urry (2000). The city and the car. *International Journal of Urban and Regional Research*, 24, 737–757.

Sheller, M. & J. Urry (2006). The new mobilities paradigm. *Environment and Planning A*, 38, 207–226.

Shin, D.-H. & F. Biocca (2017). Health experience model of personal informatics: the case of a quantified self. *Computers in Human Behavior*, 69, 62–74.

Shin, D. W., H.-K. Joh, J. M. Yun, H. T. Kwon, H. Lee, H. Min, J.-H. Shin, W. J. Chung, J. H. Park & B. Cho (2016). Design and baseline characteristics of participants in the Enhancing Physical Activity and Reducing Obesity through Smartcare and Financial Incentives (EPAROSFI): a pilot randomized controlled trial. *Contemporary Clinical Trials*, 47, 115–122.

Shoval, N., Y. Schvimer & M. Tamir (2018). Tracking technologies and urban analysis: adding the emotional dimension. *Cities*, 72, 34–42.

Sieber, R. (2006). Public participation geographic information systems: a literature review and framework. *Annals of the Association of American Geographers*, 96, 491–507.

Simonite, T. (2018). When it comes to gorillas, Google photos remains blind. https://www.wired.com/story/when-it-comes-to-gorillas-google-photos-remains-blind/, accessed 8 June 2018.

Skwarek, M. (2018). Augmented Reality activism. In *Augmented Reality art: from an emerging technology to a novel creative medium*. Second edition, ed. V. Geroimenko, 3–40. Springer, Cham.

Smith, M., C. Szongott, B. Henne & G. v. Voigt (2012). Big data privacy issues in public social media. In 2012 6th IEEE international conference on digital ecosystems and technologies (DEST), 1–6. https://doi.org/10.1109/DEST.2012.6227909, accessed 27 January 2020.

South Wales Police (2018). Facial recognition. https://www.south-wales.police.uk/en/advice/facial-recognition-technology/, accessed 8 June 2016.

Spinney, J. (2011). A chance to catch a breath: using mobile video ethnography in cycling research. *Mobilities*, 6, 161–182.

Sprenger, P. (1999). Sun on privacy: 'get over it'. https://www.wired.com/1999/01/sun-on-privacy-get-over-it/, accessed 5 June 2018.

Springhall, J. (1998). Censoring Hollywood: youth, moral panic and crime/gangster movies of the 1930s. *The Journal of Popular Culture*, 32, 135–154.

Statham, N. (2018). Use of photogrammetry in video games: a historical overview. *Games and Culture*, 1–19. doi:10.1177/1555412018786415

Stone, R. (2018). What lies beneath: virtual and augmented reality techniques for maritime heritage. In *Underwater worlds: submerged visions in science and culture*, ed. W. Abberley, 208–238. Cambridge Scholars Publishing, Newcastle upon Tyne.

Striphas, T. (2015). Algorithmic culture. *European Journal of Cultural Studies*, 18, 395–412.

Strohm, K. (2012). When anthropology meets contemporary art: notes for a politics of collaboration. *Collaborative Anthropologies*, 5, 98–124.

Sudnow, D. (1983). *Pilgrim in the microworld*. Heinemann, London.

Sun, Y., Y. Du, Y. Wang & L. Zhuang (2017). Examining associations of environmental characteristics with recreational cycling behaviour by street-level Strava data. *International Journal of Environmental Research and Public Health*, 14, 644.

Swan, M. (2012). Health 2050: the realization of personalized medicine through crowdsourcing, the quantified self, and the participatory biocitizen. *Journal of Personalized Medicine*, 2, 93–118.

Takhvar, M. (1988). Play and theories of play: a review of the literature. *Early Child Development and Care*, 39, 221–244.

Tanczer, L., T. Patel, S. Parkin & G. Danezis (2018). Gender and IoT (G-IoT) resource list. https://www.ucl.ac.uk/steapp/research/projects/digital-policy-lab/g-iot-resource-list, accessed 10 August 2018.

Tateno, M., N. Skokauskas, T. A. Kato, A. R. Teo & A. P. S. Guerrero (2016). New game software (Pokémon Go) may help youth with severe social withdrawal, hikikomori. *Psychiatry Research*, 246, 848–849.

Tatler, B. W., R. G. Macdonald, T. Hamling & C. Richardson (2016). Looking at domestic textiles: an eye-tracking experiment analysing influences on viewing behaviour at Owlpen Manor. *Textile History*, 47, 94–118.

Taylor, N. G., L. Hagen, E. Dincelli & K. Unsworth (2017). Wearable devices: information privacy, policy and user behavior. *Proceedings of the Association for Information Science and Technology*, 54, 603–605.

Thaler, R. & C. Sunstein (2008). *Nudge: improving decisions about health, wealth and happiness*. Yale University Press, New Haven.

The Conservative Party (2008). *Giving the public a crime map: using technology to fight crime*. The Conservative Party, London.

The Economist (2017). The world's most valuable resource is no longer oil, but data. https://www.economist.com/leaders/2017/05/06/the-worlds-most-valuable-r esource-is-no-longer-oil-but-data, accessed 28 August 2019.

The Warwick Commission (2015). *Enriching Britain: culture, creativity and growth. The 2015 Report by the Warwick Commission on the future of cultural value.* University of Warwick, Warwick.

Tien, T., P. H. Pucher, M. H. Sodergren, K. Sriskandarajah, G.-Z. Yang & A. Darzi (2014). Eye tracking for skills assessment and training: a systematic review. *Journal of Surgical Research*, 191, 169–178.

Tiggemann, M. & A. Slater (2013). NetGirls: the internet, Facebook, and body image concern in adolescent girls. *International Journal of Eating Disorders*, 46, 630–633.

Tiggemann, M. & M. Zaccardo (2016). 'Strong is the new skinny': a content analysis of #fitspiration images on Instagram. *Journal of Health Psychology*, 23, 1003–1011.

Tixier, N. (2016). Le transect urbain. Pour une écriture corrélée des ambiances et de l'environnement. In *Ecologie urbaines: sur le terrain*, eds. S. Barles & N. Blanc, 130–148. Economica.

Tobii (2019). Tobii Pro VR analytics: user's manual. https://www.tobiipro.com/sitea ssets/tobii-pro/user-manuals/tobii-pro-vr-analytics-user-manual.pdf/?v=1.2, accessed 20 June 2019.

Todd, C. (2012). 'Troubling' gender in virtual gaming spaces. *New Zealand Geographer*, 68, 101–110.

Tomkys Valteri, E. (2018). Presentation given at the 2nd Annual Digital Geographies Working Group symposium, Sheffield, 6 July 2018.

Udesky, L. (2010). The ethics of direct-to-consumer genetic testing. *The Lancet*, 376, 1377–1378.

Ulrich, R. (1984). View through a window may influence recovery from surgery. *Science*, 224, 420–421.

UNESCO (2019). *I'd blush if I could: closing gender divides in digital skills through education.* Equals Global Partnership, Berlin.

Vaniea, K., E. Tallyn & C. Speed (2017). Capturing the connections: unboxing Internet of Things devices. https://arxiv.org/ftp/arxiv/papers/1708/1708.00076.pdf, accessed 10 August 2018.

Vannini, P. (2016). Low and slow. https://vimeo.com/172912004, accessed 13 August 2019.

Vannini, P. (2017). Low and slow: notes on the production and distribution of a mobile video ethnography. *Mobilities*, 12, 155–166.

Vansteenkiste, P., L. Zeuwts, G. Cardon, R. Philippaerts & M. Lenoir (2014). The implications of low quality bicycle paths on gaze behavior of cyclists: a field test. *Transportation Research Part F: Traffic Psychology and Behaviour*, 23, 81–87.

Vansteenkiste, P., L. Zeuwts, M. van Maarseveen, G. Cardon, G. Savelsbergh & M. Lenoir (2017). The implications of low quality bicycle paths on the gaze behaviour of young learner cyclists. *Transportation Research Part F: Traffic Psychology and Behaviour*, 48, 52–60.

Vargas-Iglesias, J. J. & L. Navarrete-Cardero (2019). Beyond rules and mechanics: a different approach for ludology. *Games and Culture*, 1–22. doi:10.1177/1555412018822937

Vassallo, T., E. Levy, M. Madansky, H. Mickell, B. Porter, M. Leas & J. Oberweis (2016). Elephant in the valley. https://www.elephantinthevalley.com/, accessed 26 August 2019.

Vaughan-Nichols, S. (2017). How to keep your smart TV from spying on you. https://www.zdnet.com/article/how-to-keep-your-smart-tv-from-spying-on-you/, accessed 10 August 2018.

Vaughan, L., D. L. C. Clark, O. Sahbaz & M. Haklay (2005). Space and exclusion: does urban morphology play a part in social deprivation? *Area*, 37, 402–412.

Vergunst, J. (2011). Technology and technique in a useful ethnography of movement. *Mobilities*, 6, 203–219.

Vitores, A. & A. Gil-Juárez (2016). The trouble with 'women in computing': a critical examination of the deployment of research on the gender gap in computer science. *Journal of Gender Studies*, 25, 666–680.

Volkova, S., Y. Bachrach, M. Armstrong & V. Sharma (2015). Inferring latent user properties from texts published in social media. Proceedings of the Twenty-ninth AAAI conference on artificial intelligence, 4296–4297. https://www.aaai.org/ocs/index.php/AAAI/AAAI15/paper/view/9358, accessed 27 January 2020.

Vos, P., P. De Cock, V. Munde, K. Petry, W. Van Den Noortgate & B. Maes (2012). The tell-tale: what do heart rate; skin temperature and skin conductance reveal about emotions of people with severe and profound intellectual disabilities? *Research in Developmental Disabilities*, 33, 1117–1127.

Wachowski Siblings (1999). *The Matrix*. Warner Brothers.

Wachter-Boettcher, S. (2017). *Technically wrong: sexist apps, biased algorithms, and other threats of toxic tech*. W.W. Norton, New York.

Wagner, P. J. & P. Curran (1984). Health beliefs and physician identified "worried well". *Health Psychology*, 3, 459–474.

Wakefield, A. J., S. H. Murch, A. Anthony, J. Linnell, D. M. Casson, M. Malik, M. Berelowitz, A. P. Dhillon, M. A. Thomson, P. Harvey, A. Valentine, S. E. Davies & J. A. Walker-Smith (1998). RETRACTED: ileal-lymphoid-nodular hyperplasia, non-specific colitis, and pervasive developmental disorder in children. *The Lancet*, 351, 637–641.

Walker, P. (2016). City planners tap into wealth of cycling data from Strava tracking app. https://www.theguardian.com/lifeandstyle/2016/may/09/city-planners-cycling-data-strava-tracking-app, accessed 3 July 2018.

Wang, H., D. Can, A. Kazemzadeh, F. Bar & S. Narayanan (2012). A system for real-time Twitter sentiment analysis of 2012 U.S. presidential election cycle. In *Proceedings of the ACL 2012 system demonstrations*, 115–120. Association for Computational Linguistics, Jeju Island, Korea.

Wang, S. (2018). Calculating dating goals: data gaming and algorithmic sociality on Blued, a Chinese gay dating app. *Information, Communication & Society*, 1–17. doi:10.1080/1369118X.2018.1490796

Warren, S. (2017). Pluralising the walking interview: researching (im)mobilities with Muslim women. *Social & Cultural Geography*, 18, 786–807.

Warren, S. D. & L. D. Brandeis (1890). The right to privacy. *Harvard Law Review*, 4, 193–220.

Watts, L. & J. Urry (2008). Moving methods, travelling times. *Environment and Planning D: Society and Space*, 26, 860–874.

Watts, S. & P. Stenner (2012). *Doing Q methodological research: theory, method and interpretation*. Sage, London.

Webster, A. (2015). Soon you'll be able to control Assassin's Creed with your eyes. https://www.theverge.com/2015/2/5/7986565/assassins-creed-rogue-tobii-eye-tracking, accessed 18 June 2019.

Webster, A. (2019). Building a better Paris in Assassin's Creed Unity. https://www.the verge.com/2014/10/31/7132587/assassins-creed-unity-paris, accessed 26 April 2019.

Weinstein, J., W. Drake & N. Silverman (2015). Privacy vs. public safety: prosecuting and defending criminal cases in the post-Snowden era. *American Criminal Law Review*, 52, 729–752.

Westin, A. (1967). *Privacy and freedom*. Atheneum, New York.

Whitson, J. R. (2018). What can we learn from studio studies ethnographies? A "messy" account of game development materiality, learning, and expertise. *Games and Culture*, 1–23. doi:10.1177/1555412018783320

Whyte, W. H. (1980). *The social life of small urban spaces*. Project for Public Spaces, New York.

Wilken, R. (2010). A community of strangers? Mobile media, art, tactility and urban encounters with the other. *Mobilities*, 5, 449–468.

Wilmott, C. (2016a). In-between mobile maps and media: movement. *Television & New Media*, 18, 320–335.

Wilmott, C. (2016b). Small moments in spatial big data: calculability, authority and interoperability in everyday mobile mapping. *Big Data & Society*, 3, 1–16.

Windsor, R. (2015). Riders urged to consider Strava privacy settings after thieves target Manchester cyclist. www.cyclingweekly.com/news/latest-news/riders-urged-to-consider-strava-privacy-settings-after-thieves-target-manchester-cyclist-205098, accessed 3 July 2018.

Witkowski, E. (2018a). Running with zombies: capturing new worlds through movement and visibility practices with Zombies, Run! *Games and Culture*, 13, 153–173.

Witkowski, E. (2018b). Sensuous proximity in research methods with expert teams, media sports, and esports practices. *MedieKultur: Journal of Media and Communication Research*, 34, 31–51.

Witkowski, E. & J. Manning (2019). Player power: networked careers in esports and high-performance game livestreaming practices. *Convergence*, 25, 953–969.

Wolf, G. (2016). The quantified self: reverse engineering. In *Quantified: biosensing technologies in everyday life*, ed. D. Nafus, 67–72. MIT Press, Cambridge MA.

Wood, D. & J. Fels (1993). *The power of maps*. Routledge, London.

Woodley, S. (2019). Boy sentenced to 18 years in prison following Selly Oak stabbing. https://www.redbrick.me/selly-oak-stabbing-sentence/, accessed 14 August 2019.

Wright, R., L. John, R. Alaggia & J. Sheel (2006). Community-based arts program for youth in low-income communities: a multi-method evaluation. *Child and Adolescent Social Work Journal*, 23, 635–652.

Yeung, K. (2018). Algorithmic regulation: a critical interrogation. *Regulation & Governance*, 12, 505–523.

Yip, N. M., R. Forrest & S. Xian (2016). Exploring segregation and mobilities: application of an activity tracking app on mobile phone. *Cities*, 59, 156–163.

Zarsky, T. (2016). The trouble with algorithmic decisions: an analytic road map to examine efficiency and fairness in automated and opaque decision making. *Science, Technology, & Human Values*, 41, 118–132.

Zeile, P., B. Resch, J.-P. Exner & G. Sagl (2015). Urban emotions: benefits and risks in using human sensory assessment for the extraction of contextual emotion information in urban planning. In *Planning support systems and smart cities*, eds. S. Geertman, J. Ferreira, R. Goodspeed & J. C. H. Stillwell, 209–225. Springer, Berlin.

Zendle, D., P. Cairns & D. Kudenko (2015). Higher graphical fidelity decreases players' access to aggressive concepts in violent video games. In *Proceedings of the 2015 Annual symposium on computer-human interaction in play*, 241–251. ACM, London.

Zeuwts, L., P. Vansteenkiste, F. Deconinck, M. van Maarseveen, G. Savelsbergh, G. Cardon & M. Lenoir (2016). Is gaze behaviour in a laboratory context similar to that in real-life? A study in bicyclists. *Transportation Research Part F: Traffic Psychology and Behaviour*, 43, 131–140.

Zoellner, M., J. Keil, T. Drevensek & H. Wuest (2009). Cultural heritage layers: integrating historic media in augmented reality. In *2009 15th International conference on virtual systems and multimedia*, 193–196. https://doi.org/10.1109/VSMM.2009.35, accessed 27 January 2020.

Zook, M. A. & M. Graham (2007). Mapping DigiPlace: geocoded internet data and the representation of place. *Environment and Planning B-Planning & Design*, 34, 466–482.

Zraic, K. (2019). Sex toy award is restored by trade show after an outcry over sexism. https://www.nytimes.com/2019/05/09/technology/sex-toy-award-vibrator.html, accessed 30 May 2019.

Index

Printed in the United States
by Baker & Taylor Publisher Services

Printed in the United States
by Baker & Taylor Publisher Services